# The Ecstatic Journey

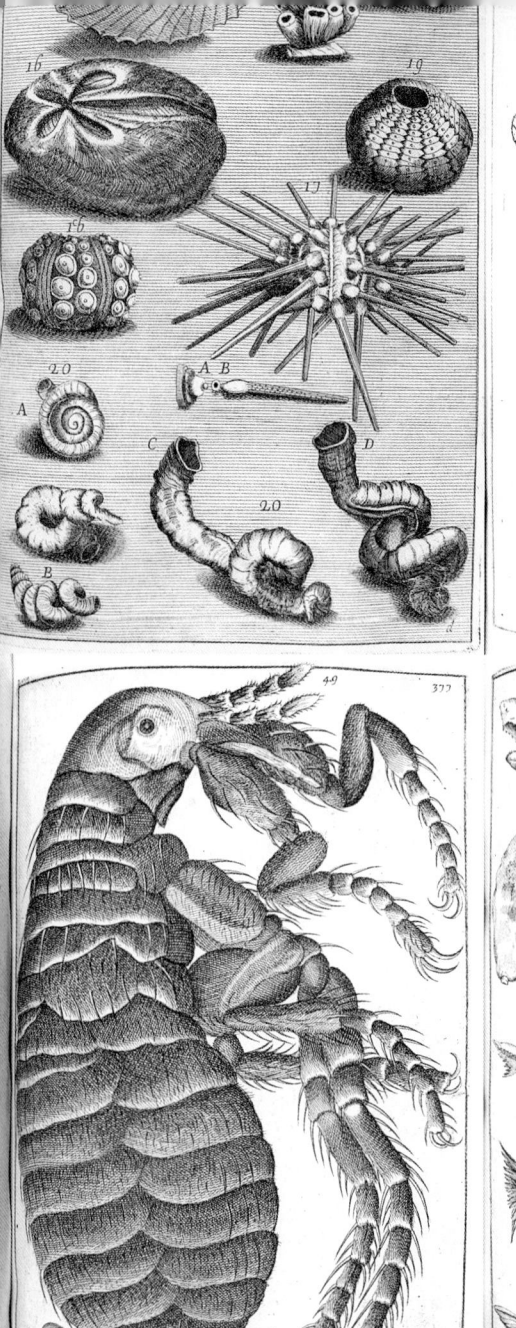
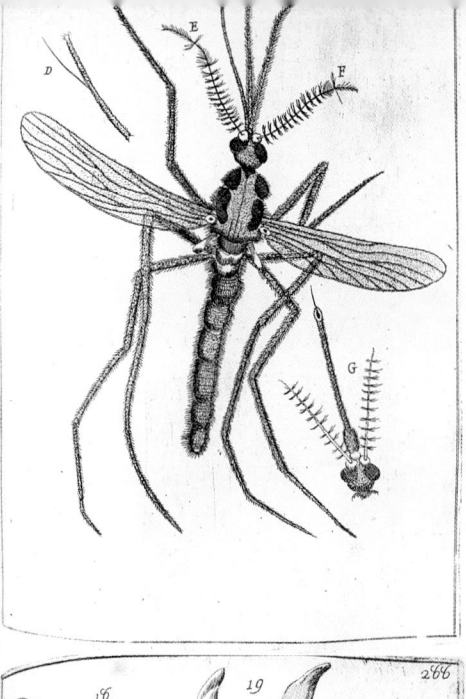
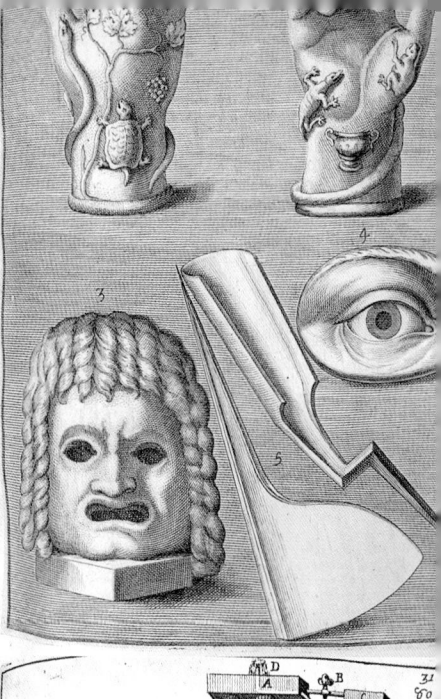

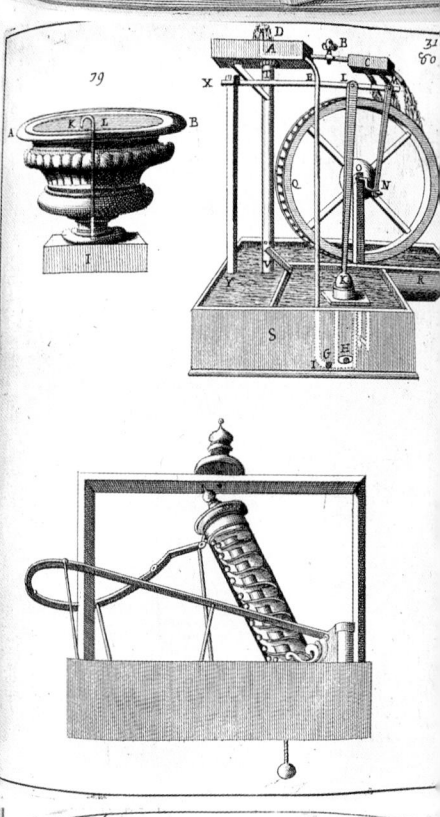
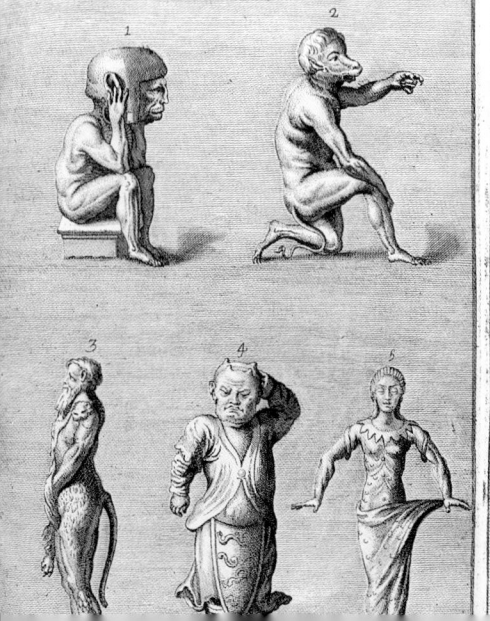
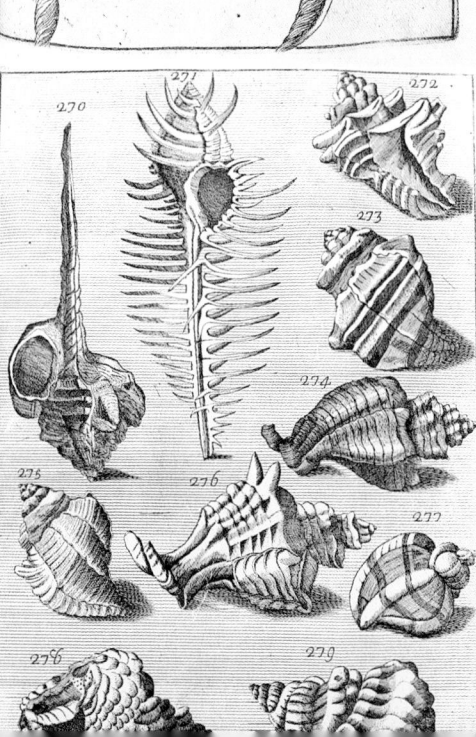
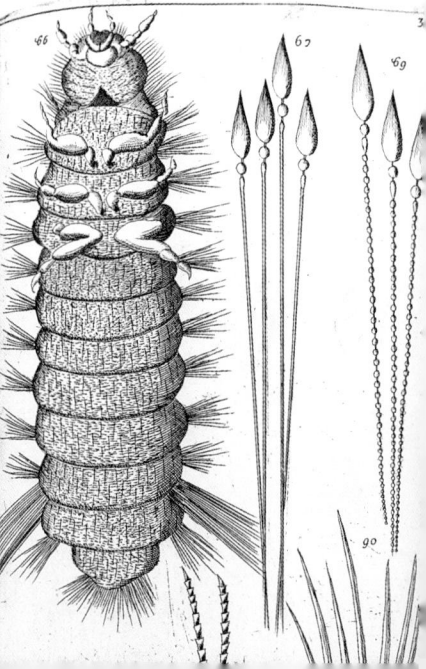

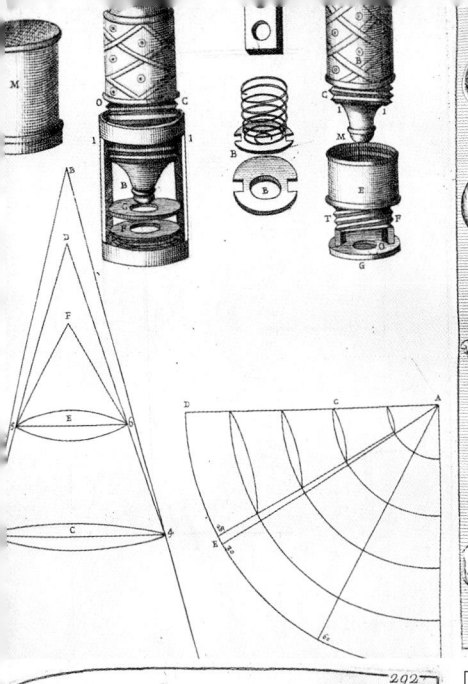

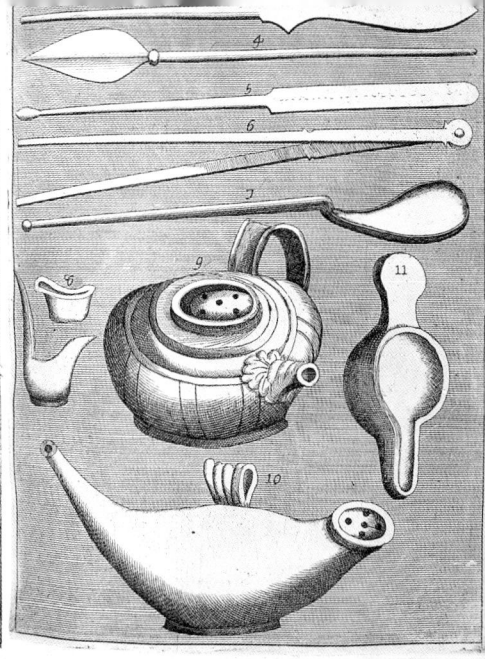

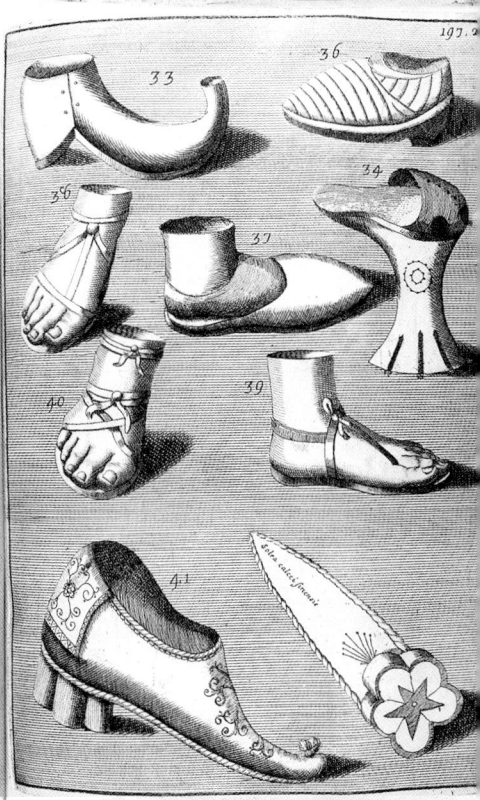

# The Ecstatic Journey: Athanasius Kircher in Baroque Rome

BY

Ingrid D. Rowland

WITH AN INTRODUCTION BY

F. Sherwood Rowland

UNIVERSITY OF CHICAGO LIBRARY
2000

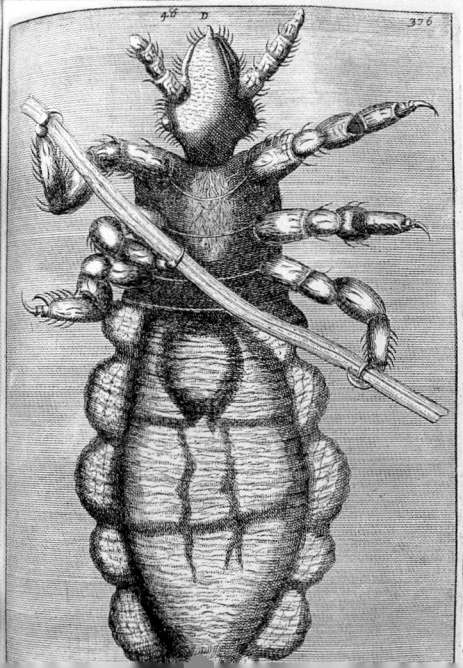

1,750 copies of this catalogue were published in conjunction with an exhibition held in the Department of Special Collections University of Chicago Library February 1–April 7, 2000.

Support for this publication was provided by the University of Chicago Library Society.

©2000 University of Chicago
All rights reserved.
For permission to quote or reproduce from this catalogue contact:
University of Chicago Library
1100 East 57th Street
Chicago, Illinois 60637
*www.lib.uchicago.edu/e/spcl/*

Library of Congress Cataloging-in-Publication Data

Rowland, Ingrid D. (Ingrid Drake)
    The ecstatic journey: Athanasius Kircher in Baroque Rome/ by Ingrid D. Rowland; with an introduction by F. Sherwood Rowland.
        p. cm.
    Includes bibliographical references.
    ISBN 0-943056-25-x
        1. Science–Italy–Rome–History–17th century–Exhibitions. 2. Kircher, Athanasius, 1602-1680–Exhibitions. 3. Rome (Italy)–Intellectual life–17th century–Exhibitions. I. University of Chicago. Library. Dept. of Special Collections. II. Title.
Q127.I8 R68 2000
509.45'63'09032–dc21                               00-27481

Design and typesetting by Joan Sommers Design.
Copy-editing by Britt Salvesen.
Photography by Stephen Longmire and Ted Lacey.
Produced by the Department of Special Collections, University of Chicago Library.
Printed by C & C Offset Printing Co., Ltd., Hong Kong.

Images on the covers and title page are from the following catalogue items: 1, 4 (title page and front cover); 2 (front cover); 5, 9 (back cover); 44, 69, 75, 79, and 93 (front cover).

*Note:* Translations are all by Ingrid D. Rowland unless otherwise noted. Biblical references are to the Authorized (King James) version. Titles and imprints in bibliographic citations are transcribed from the texts; elsewhere, standard spellings of titles are used.

# Contents

Preface
ALICE SCHREYER
vii

INTRODUCTION
Looking Back From the Twenty-first Century:
Athanasius Kircher and the Beginnings of Science
F. SHERWOOD ROWLAND
ix

ESSAY
Athanasius Kircher, Missionary Scientist
INGRID D. ROWLAND
1

EXHIBITION TEXT
The Ecstatic Journey:
Athanasius Kircher in Baroque Rome
31

# Preface

"The Ecstatic Journey: Athanasius Kircher in Baroque Rome" is the first full-scale exhibition presented by the Department of Special Collections following an extensive reconfiguration project at the Joseph Regenstein Library. We are delighted to mark the occasion with this exciting project and grateful to Professor Ingrid Rowland for conceptualizing and realizing it.

The exhibition surveys the scientific, religious, and political culture of seventeenth-century Rome through Athanasius Kircher's amazing world of magic lanterns, volcanoes, fossils, flying cats, hieroglyphics, and practical jokes with the most serious of intentions. Many landmark scientific works are included, alongside archaeological, philological, classical, and religious treatises magnificent and modest, renowned and obscure. Professor Rowland's introductory essay presents an overview that is expanded and developed in the complete exhibition text and detailed descriptions of individual items that follow. The extraordinary depth of seventeenth-century holdings at the University of Chicago Library makes it possible to show several key titles in more than one edition or copy, illustrating the range of topics encompassed by Kircher's writings. Translations and quotations from other works and writers provide contextual information.

As "The Ecstatic Journey" demonstrates, although many of Kircher's discoveries have been discounted, his commitment to the experimental method and his relentless curiosity are of permanent value and continuing fascination. Also beyond dispute is the extent to which he promoted and promulgated his ideas through forty-plus books, exploiting the highly sophisticated printing industry of seventeenth-century Europe. "The Ecstatic Journey" thus sheds light on several aspects of the book trade in the Baroque era, including the production of monumental illustrated books; the availability of exotic type fonts; licensing and censorship; the influence of religion, politics, and patronage on the production and distribution of printed works.

The idea for an exhibition devoted to the intellectual life and circle of Athanasius Kircher arose when Professor Rowland introduced him in a talk to the University of Chicago Library Society in October 1998. Professor Rowland was already planning a Smart Museum exhibition entitled "The Place of the Antique in Early Modern Europe," the culmination of a graduate seminar organized under the auspices of a grant from the Andrew W. Mellon Foundation. The strength of the Library's collections in works by Kircher and his contemporaries sparked interest in a parallel project investigating the impact of religion on Kircher's science. During winter 1999, Professor Rowland offered a seminar on Baroque Rome in which she explored some themes developed in the exhibition.

Early on, assessing Kircher's contributions to the spirit of scientific inquiry emerged as a principal aim of the project. In this context, Professor Rowland described conversations about Kircher with her father, F. Sherwood Rowland (SM'51, PhD'52, Hon DSc'89, and recipient of the University's 1997 Alumni Medal). Professor Sherry Rowland, 1995 Nobel laureate for his work along with two other scientists on the destructive effects of chlorofluorocarbons, or CFCs, on the earth's ozone layer, generously agreed to write an introduction to this catalogue. His essay is an elegant appreciation of Kircher and a meditation on the concept of scientific progress.

In the course of putting the exhibition and publication together, Professor Rowland proved herself to be as indefatigable as Father Kircher himself. During the research phase, construction noise and disruptions were a constant feature in the Department of Special Collections, and Professor Rowland graciously ignored them all. Text came forth on schedule from Rome and the hills above Los Angeles, where Professor Rowland was a scholar in residence at the Getty Research Institute during the 1999 fall quarter. Her unfailing enthusiasm and good humor filled all who worked on the exhibition with great admiration for both Kircher and his late twentieth-century advocate.

Many other colleagues within and outside the Library made essential contributions to this exhibition and publication. The University of Chicago Library Society provided generous funding for the catalogue. Art history graduate student Mario Pereira assisted with early research investigations. Kimerly Rorschach, Dana Feitler Director, Smart Museum of Art, strongly supported a collaborative approach to the Library and Smart Museum projects; and Coordinating Curator for Mellon Projects Elizabeth Rodini assisted with a number of details. Within the Department of Special Collections, Barbara Gilbert, Debra Levine, Jay Satterfield, and Jessica Westphal facilitated the examination of many volumes during a period of severe constraints on our services. Valerie Harris provided meticulous bibliographic support, with contributions by Deborah Derylak and Susanna Morrill. Valarie Brocato skillfully and gracefully managed innumerable details surrounding the production of the exhibition and catalogue. She developed a striking exhibition design that did full justice to Kircher's world and the works on view, and, with assistance from Theresa Smith, expertly produced the show. "The Ecstatic Journey: Athanasius Kircher in Baroque Rome" illustrates how the richness of the Library's collections, and the staff who work with researchers and materials, support the process of scholarly discovery, interpretation, and presentation at the University of Chicago.

ALICE SCHREYER
*Curator, Department of Special Collections*

INTRODUCTION

# Looking Back From the Twenty-first Century: Athanasius Kircher and the Beginnings of Science

JUST OUT OF GRADUATE SCHOOL in the 1950s, I briefly considered studying the chemical processes involved in the emission of visible light from the reaction with the oxygen in air of a chemical known as luciferin. This glow has been seen clearly at close range by billions of people over thousands of years, although most of their descriptions are very different from mine above: usually much more excited, and often poetic. Everyone who has ever been fascinated by the twinkling lights of fireflies in the early evening was encountering this same luciferin-oxygen reaction in its natural surroundings, and millions have probably thought to themselves as they watched, "I wonder how they do that?" Mankind's curiosity about such observable natural phenomena has served as the basic driving force leading to most advances in science. Indeed, scientific studies were described as natural philosophy until the late nineteenth century, when expanding knowledge resulted in the division into the individual sciences such as physics, chemistry, and biology.

As this century ends, a scientist pondering fireflies can go directly to the scientific literature and readily find that the luciferin molecule, whose name comes from the Latin, *lucifer*, meaning "bearer of light," contains eleven atoms of carbon, eight of hydrogen, three of oxygen, and two each of nitrogen and sulfur, arranged in space and connected to one another in a structure whose order and dimensions are all known with high precision. Most of this information is the product of scientific studies carried out during the last half century, while the chemical isolation (and naming) of a pure sample of luciferin by the French physiologist Raphael Dubois took place just over a century ago. As one goes still further back into history, the firefly explanations become less quantitative, less certain, and more speculative, until you come to the first person who recorded his explanation of the visible flashes in the night, Athanasius Kircher, a German Jesuit priest living in Rome in the seventeenth century. While his conclusions are no longer cited as authoritative, his explanation of the firefly glow began the trail that led to our current understanding of the process. In modern technical terminology, Kircher is credited with initiating the study of bioluminescence, the emission of visible light by living creatures.

The modern scientist curious about this phenomenon can turn successively to encyclopedias, organic chemistry textbooks, monographs on bioluminescence, and finally to many thousands of individual scientific papers in search of ever-deepening understanding. These volumes record the structure of luciferin and that of its necessary enzyme companion, luciferase, which positions the oxygen molecules in careful juxtaposition to luciferin in order for the chemical reaction to occur. The particular mixture of red, yellow, and blue wavelengths of light that together appear whitish in the evening gloom is known in detail. The chemical differences between firefly luciferin and several dissimilar, yet also light-producing, molecules found in marine crustaceans and other biological species are fully established. However, unlike Kircher, whose interest stemmed directly from observations of nature, the scientist of the year 2000 A.D. is not very likely to approach this accumulated scientific knowledge from speculation about fireflies. Instead, he or she is much more likely to be responding to some recondite aspect of nature hidden from direct observation by human eyes—perhaps the structural contrasts and similarities between luciferase and the quite different enzyme which facilitates the glow from the sea pansy.

*Kircher is credited with initiating the study of bioluminescence, the emission of visible light by living creatures.*

Athanasius Kircher had no such trove of scientific knowledge in the 1650s. Most of the information we have was as yet undiscovered, and therefore unavailable, to Dubois only a century ago. Viewed retrospectively from a vantage point 350 years into the future, Kircher had access to essentially zero in scientific information. However, the screen was not entirely blank—the learned people of the sixteenth century had large amounts of accumulated and organized wisdom. Because science as we now know it was in its infancy, this wisdom was codified, evaluated, judged, and guarded by the various religious establishments. While some of this sixteenth- and seventeenth-century lore is still accepted in the twentieth century, for instance the theory that pieces of iron and stone could arrive from outside the earth's atmosphere; some is not—for example, the belief that the sun revolves around the stationary earth. The transition from the seventeenth century to the threshold of the twenty-first has not always been smooth; there have been some reversals along the way. During the eighteenth century, the concept of meteorites was rejected because in an age of reason it seemed obvious to every intelligent person that stones do not fall from the sky. Rehabilitation of the meteorite concept came early in the nineteenth century through a succession of eyewitness reports, plus the physical existence—in sharp contrast to the twentieth-century sightings of UFOs—of misshapen chunks of iron unlike any other iron-containing earth objects. The majority of known meteorites has been found in the past thirty years in Antarctica, where both iron and stone objects are visibly foreign in the barren, windswept, nearly ice-free deserts.

The present-day scientist is surrounded by the same natural phenomena as was Kircher, with the great difference that now most of the basic details of these phenomena are reasonably well understood. For example, common table salt consists of equal numbers of sodium atoms and chlorine atoms; one of the eleven electrons of a neutral sodium atom has been added to the seventeen electrons already present in a neutral chlorine atom to form a positive sodium ion and a negative chlorine ion attracted to each other by the opposite charges. The evidence is so overwhelming that it is no longer even cited. In our day, the scientist accepts that the great bulk of scientific lore is correct, or at least sufficiently close to correct, in order to get to the exciting fringe of knowledge where major unknowns can be encountered. The scientist moving out toward the edges of knowledge is looking for something amiss—something that doesn't seem quite complete or right. He or she then aims an experiment at that perhaps wobbly construct. Kircher, in contrast, had almost complete freedom of choice in subject matter—in his view, essentially everything needed explanation. And explain he did, in more than forty works.

Kircher realized that if the world were only 6,000 years old as interpreted from the Bible, rapid change must be a characteristic of the cosmos. In the present day, the concept of a 6,000-year-old earth remains with us, although not in scientific circles, and is still an arena of controversy. However, the questions of global change in a single century and biological evolution over hundreds of millennia are under daily discussion both within science and in the general public. Of course, theories of evolution and the 6,000-year-old earth are frequently interconnected as religious beliefs confront modern scientific conclusions.

Although wisdom is highly valued by societies, challenges to accepted wisdom are usually not appreciated—indeed, the more significant a belief is to the society, the greater the hostility toward any questioning of its correctness. Furthermore, in Kircher's time as at present, those who disagree with the established order pay a high price. In the seventeenth century, the nascent sciences collided with wisdom authorized by religious certainty; and the penalties suffered by Kircher's predecessors Galileo Galilei and Giordano Bruno included imprisonment and being burned at the stake. Twentieth-century penalties, at least in the Western world, are trifling in comparison, probably because most of the displaced wisdom comes from within the scientific community itself. But the controversies are ever-present between new insights and the accepted world view. More than one twentieth-century scientist has chosen not to debate loudly in opposition to a state-supported view (as, for example, was the case when Lysenko's theories of genetics dominated Soviet biology), even without the threat of one's funeral pyre as an imminent corollary. Could Kircher have sometimes been

> *Kircher realized that if the world were only 6,000 years old as interpreted from the Bible, rapid change must be a characteristic of the cosmos.*

prudent rather than argumentative about concepts—perhaps of lesser importance to him—with which he did not fully agree?

Those of us who have lived through the last half century of science are keenly aware of the importance that advances in instrumentation play in aiding more complete understanding of particular phenomena. One major advantage is that an instrument records some phenomenon with much greater precision than human capability (e.g. the stopwatch, and, earlier in history, the clock itself). The other is the ability to sense some characteristic associated with the phenomena that are not readily detectable by human senses (e.g. ultraviolet radiation, to which the human eye does not respond). In the last few decades, we have had instruments that perform both functions. Beginning in the late sixteenth century, improvements in optics allowed the telescope of Galileo to see distant planets and moons more clearly, and the microscope of Anton van Leeuwenhoek to see objects too small for the unaided human eye. Kircher, too, experimented with a rudimentary microscope in his efforts to understand phenomena in more detail. As instruments became more important, so did the ability to operate and maintain them; and specialization became a central characteristic of advances in science. As quantitative measurements gradually displaced qualitative observations, reproducible instrument performance required stable external conditions. Indoor laboratories became indispensable, followed by climate control within those laboratories. This natural progression, however, brought with it loss of interest in the related natural phenomena going on outside the laboratory. Kircher, with much to explain to everyone, was concerned not only with the facts themselves, but with their presentation to the neophyte. It is perhaps not surprising that he is recognized today as an early user of the magic lantern, the forerunner of today's slide and transparency projection systems.

Difficulty often attends attempts to understand scientific efforts of the past—in effect, to imagine oneself in the place of earlier scientists with only the available knowledge of the time. "How did he [then it was almost always he] reach that conclusion without knowing x, y, and z, which were discovered 50 and 100 years later?" One result of this exercise has always been to make me realize the very great talents of some of the scientists of 100 or 200 years ago. I especially admire persons such as Kircher, who took on problems across a very wide range of fields as we know them, not with uniform success but with great skill in putting together plausible explanations for diverse phenomena in physics, chemistry, and biology. Such retrospective ruminations can also be very interesting and informative when an investigation went a long way toward the correct solution, but never quite found the key. One of Kircher's contributions concerns the decipherment of Egyptian hieroglyphics—an intriguing problem for more than a millennium after the ability to write in hieroglyphs died out. Every scientist can imagine the effect of an analog of the Rosetta Stone on one's own particular scientific inquiry, when the key to breaking the code suddenly becomes available and the possibility of success can be envisioned, even though much hard work often remains. Kircher himself approached the Egyptian hieroglyphics with correct assumptions that the symbols were phonetic signs and that they were an earlier stage of the Coptic language, but then he went astray. One can only speculate about what would have happened if the real Rosetta Stone had been discovered 150 years earlier, in his time rather than Napoleon's, and taken to Kircher for interpretation.

Fast-forwarding through several centuries of scientific development, the minute details required for understanding the physics or chemistry of a particular system came to be less and less related to the external natural phenomena that were the initial subjects for inquiry. The descriptive "natural philosophy" of the mid-nineteenth century was fragmented into chemistry, physics, astronomy, biology, meteorology, and then into the composites—biophysics, sociobiology, and my own atmospheric chemistry—of the present; and scientists became more and more specialists in their own deeper, narrower areas. Now, when attention returns to the natural phenomena outside the laboratory, the complexity of the real world rarely puts the problem squarely in the lap of a single science, and interdisciplinary investigation has become the norm. The distance from Athanasius Kircher to the twenty-first century is a long one, but the intellectual and scientific thread is there.

F. SHERWOOD ROWLAND
*Donald Bren Research Professor of Chemistry
and Earth System Science
University of California, Irvine*

OBELISCI IN AREA AEDIS S MACHVTI ROMAE SITI QVADRIPARTITA DELINEATIO

CAT. 87

ESSAY

# Athanasius Kircher, Missionary Scientist

WHEN THE GERMAN JESUIT Athanasius Kircher arrived in Rome in 1635, his reputation had long preceded him: among the twelve languages he claimed to command he included—uniquely for his time—the ability to read ancient Egyptian hieroglyphs. He also constructed mechanical devices of marvelous ingenuity, conducted scientific experiments, and seemed to know new and exciting information about virtually every subject under the sun, whose spots and firestorms he had observed with glee through his own telescope.

Officially Father Kircher took up the chair in mathematics at the Jesuit Order's Roman College, the Collegio Romano, an imposing complex built over the ruins of the ancient Roman temple of Isis. It was a strikingly appropriate setting for the world's acknowledged master of hieroglyphics. Furthermore, an injunction from the powerful Cardinal Francesco Barberini, the intellectually inclined nephew of the reigning pope, Urban VIII, granted Kircher lighter teaching duties to give him more time to prepare his studies on ancient Egypt for publication.

Kircher must have come with another mission as well, this one carefully unstated, as were so many of the missions and motives that powered seventeenth-century Rome. On June 22, 1633, Pope Urban VIII had presided over the condemnation and imprisonment of a onetime friend, Galileo Galilei, by the Roman Inquisition. The charge was "vehement suspicion of heresy," specifically stating in print and in person that the earth revolved around the sun. By the time of Kircher's arrival in Italy, the Tuscan astronomer's punishment had only recently been commuted to house arrest, and the pope was still seething. His cardinal nephew, Francesco Barberini, Kircher's mentor, had refused to sign the inquisitorial sentence (along with two other cardinals); most contemporaries who had any reason to care assumed that the Jesuit-trained Barberini's own beliefs about the structure of the cosmos favored a solar system, despite the fact that for faithful Catholics, open agreement with the writings of Copernicus had been forbidden ever since Galileo's first brush with the Inquisition in 1616.

Before the Church compelled him to silence, Galileo had been one of the great debaters of his time, and the Jesuits, as sworn defenders of the Catholic faith, had been hard-pressed to produce an opponent whose combination of scientific competence and rhetorical skill could counter the sharp-tongued, brilliant Tuscan's threat to established belief. It was no accident that most of Kircher's predecessors in the Collegio Romano's chair of mathematics had managed to spar with Galileo on one occasion or another: Orazio Grassi on comets, Christoph Scheiner on sunspots, the venerable Christoph Clavius on the revelations of the telescope (at least until Clavius came around to Galileo's way of thinking). Clavius and yet another holder of the mathematics chair, Christoph Grienberger, managed to remain on friendly terms with the prickly Galileo over the years. As a fellow Italian, Grassi could exchange invective with Mediterranean exuberance and without hard feelings. Scheiner, however, came to loathe Galileo with a physical revulsion that even the Inquisition's punishment did nothing to mitigate.

In addition to Galileo's rapier wit, all of these Jesuit colleagues contended with a deeper problem. A directive from their order compelled them to teach Aristotle's account of the universe as a finite set of concentric spheres rotating about the earth. As competent mathematicians, however, they suspected, probably to a man, that Copernicus—and Galileo after him—were right: the earth revolved around the sun. They also knew what it would cost them to say so openly.

In the troubled intellectual climate of Rome after 1633, Kircher, a flamboyant showman of cheerfully imperturbable manner, may have seemed at last to supply Catholic orthodoxy with a figure whose glamour approached Galileo's. The two would never come into direct contact: by this time, Galileo himself, old, arthritic, and nearly blind, was bound by law to silence in his Florentine villa. Nor was Kircher the ideal doctrinaire; his mind ranged too widely for that. All the same, with his Egyptian studies and his mechanical devices, Athanasius Kircher must have seemed to offer hope for some new, and perhaps less theologically fraught, topics of conversation, both within the Collegio Romano and among the gentlemen's academies of secular Rome. Besides, the course of his early career must have made it clear that unlike Grassi and Scheiner, he was not a duellist by inclination. All told, as a representative of Jesuit science (or better, natural philosophy), he must have had the look of a conciliator rather than a paladin. For Cardinal Barberini, whose identities as Jesuit alumnus, friend of Galileo, papal nephew, and probable Copernican sympathizer presented insoluble contradictions, Kircher must have promised some degree of respite from his own troubles. At Barberini's instigation, Scheiner, who had been serving the Collegio Romano as professor of mathematics, was sent to Vienna, to take up a position at the imperial court that had first been designated for Kircher. Kircher, for his part, stayed in Rome, with only one interruption, for the rest of his life.

*In the troubled intellectual climate of Rome after 1633, Kircher, a flamboyant showman of cheerfully imperturbable manner, may have seemed at last to supply Catholic orthodoxy with a figure whose glamour approached Galileo's.*

If he was not necessarily a fighter by vocation, Kircher was certainly a survivor, a very clever one at that. His autobiography recounts an astounding succession of narrow escapes, first as an overly curious child and then as a young Catholic religious in a Protestant region of Germany. Caught in a stampede of horses, washed through the workings of a water-mill (twice), marooned on a wayward ice floe, captured by Protestant mercenaries, shipwrecked with alarming frequency, Athanasius Kircher lived up to a first name that meant, in Greek, "immortal." Yet for all his gregarious manner, he could also keep to himself with iron resolve; as an aspiring Jesuit novice, out of a misplaced sense of modesty, he had initially impressed his teachers as exceptionally stupid. Then, at grave risk to his health, he concealed an excruciating case of infected chilblains for fear that the order would reject him in its search for men as physically hearty as they were spiritually staunch. Good sense finally induced him to reveal his intelligence; anguished prayer to Christ and the Virgin Mary took care of his incipient gangrene, and with mind and body set to rights, Kircher finally embarked on his career as a kind of living encyclopedia. Still, in later life, his thoughts could absorb him to the point of rapturous oblivion. As he said in 1636 to Cardinal Barberini, "I sing inwardly, to myself and to the Muses." For all his remarkable productivity as a writer, he always divulged those inward songs carefully, and only in part. It was the Jesuit way.

Already by Kircher's time, a century since the founding of their order, the men who made up the Society of Jesus had been transformed from a small band of tough-minded former soldiers into a worldwide organization with a reputation for subtlety, rhetorical agility, and shrewd sensitivity to power. The military regimen still obtained: each Jesuit was to embody within himself the whole Catholic world, and be prepared at all times to carry his personal microcosm to any region of the globe. To sustain them in this mission, Ignatius Loyola's *Spiritual Exercises* (*Exercicios espirituales*) adapted the techniques of classical memory training to specifically religious purpose, inculcating a common body of spiritual knowledge together with a formidable mental discipline, versatile enough to reach both foreign potentates and the urban poor. Each Jesuit's presumed self-sufficiency inspired the order to grant its members remarkably broad permission to adapt to local customs. Engravings from Kircher's admiring tribute to his far-flying society, *China Illustrated in Monuments* (*China monumentis illustrata*), offer striking examples of the elaborate—and idiosyncratic—Chinese dress of the missionaries Adam Schall von Bell and Matteo Ricci. Schall von Bell acted as court astronomer to the emperor in Beijing under both the Ming and Qing dynasties. Across his brocade robe, a far cry from Jesuit black, he professes his faith by wearing a scapular embroidered with a Christian symbol of Christ's self-sacrifice: a mother pelican piercing her own breast to feed her young on her own blood. Another Jesuit, Ricci, also moved, after a tumultuous initiation, within China's highest circles, teaching the art of memory to aspiring Chinese civil servants, and making a few converts on the way, like the mandarin Paul Li, with whom he poses in another image from Kircher's *China Illustrated*.

In South and East Asia, in the Americas, and in a Europe at war, Jesuit missionaries faced appalling tortures with a fortitude they traced to their soldier-founder, Ignatius; they inherited his mental agility as well. From Schall von Bell, Ricci, and the Jesuit court painter Giuseppe Castiglione in Beijing to the courts of Europe, the Society of Jesus moved with polished diplomatic skill. One of Kircher's fellow fathers in Rome, an Austrian named Melchior Inchofer, affirmed his order's sometimes sinister reputation with an insider's precision:

> [The Jesuit General] has Emissaries who are so well-versed in capturing the souls of Princes that they insinuate themselves deep into their counsels, and by anticipating information beforehand lead them where they will, most of all toward friendship with [himself], not to say obsequy and servitude. By virtue of their diligence the entire Globe would long since have obeyed the sceptre of the [Jesuits], had their unrestrained lust for domination not corrupted their successful ventures.

In the early seventeenth century, Jesuit superiors customarily requested that the reverend fathers publish their most controversial work under pseudonyms, to spare the order endless involvement in contemporary debates. Melchior Inchofer wrote under at least four: as "the Vertumnian Academic" to oppose Galileo in 1633 (when Inchofer was serving as one of his examiners on behalf of the Inquisition), as "Eugenio Lavanda" in 1640, as "Benno Durkhundurk" in 1642, and in 1645 as "Lucius Cornelius Europaeus" to pen the biting satire (quoted above) of his own order known as *The Monarchy of the Solipsists* (*Monarchia solipsorum*). Orazio Grassi sparred with Galileo as "Lothario Sarsi" and with Scheiner as "Apelles." But Athanasius Kircher always wrote in his own name, as himself, in each of the forty-odd books he published during his long career. As a crusader for Jesuit science, he preferred to work through spectacle rather than debate, playing himself in what he habitually called "the theater of the world" (*theatrum mundi*).

And the spectacles he staged in that *theatrum mundi* of Baroque Rome must have been remarkable. Kircher reveled in inventing mechanical devices as well as new words, fancifully sesquipedalian, to describe them: the *Smicroscopium*, the Maltese Observatory, the Astronomical Tube. Nor were all these machines soberly practical appliances or delicate scientific tools—for Father Kircher was also an inveterate prankster. His passage through the world could never have been inconspicuous, whether he was setting off depth charges in Lake Albano, climbing down volcanic craters in Naples and Sicily, standing under waterfalls at Tivoli, or putting food out to rot on his windowsill for later examination under the *Smicroscopium*—as he called his microscope. Two places above all felt the whirlwind impact of Kircher's presence: the Collegio Romano, where he installed the collection he called his *Musaeum*, and the tiny hamlet of Mentorella, in the Roman countryside, where he helped to restore a medieval chapel to Saint Eustace and paid the site regular devotional visits.

Eventually the peasants who lived around Mentorella must have grown accustomed to the steady stream of disruptions to the rhythm of their normal life: the hot-air balloons in the

DETAIL, CAT. 2

shape of dragons with "Flee the wrath of God" blazoned on their bellies, the irate cats dressed up in paper cherubs' wings, the wolves set to howling when one of Kircher's friends bayed at them through a megaphone, the shooting beams of light maneuvered over hill and dale by specially shaped mirrors. Sometimes the long-suffering *contadini* were themselves enlisted in Kircher's experiments, waving white handkerchiefs by day or candles by night if they heard the sound of his latest noisemaker, filing in for mass when they heard the mechanically amplified call. Kircher intended his pranks to rid these country dwellers of superstition, just as his trumpets, bells, and megaphone-enhanced voices summoned them to worship. He disliked miracles on principle, for to him Nature was already miraculous enough when she obeyed the rules.

*The Jesuits' rooms were filled with trick mirrors, barometers, thermometers,* Smicroscopia, *clocks, and dancing bronze cherubs encased in glass bulbs.*

But if Mentorella provided the unlikely scene for some extraordinary activity, life at the Collegio Romano with Kircher in residence must have been downright phantasmagoric. The site of his *Musaeum* shifted several times as his collections grew over the course of fifty years. The engraved frontispiece to its first catalogue, *The Celebrated Museum* (*Musaeum celeberrimum*), shows its latest and most spacious location, when its rooms were stuffed to bursting with Roman statues, Greek vases, Etruscan bronzes, musical instruments, wooden model obelisks, a stuffed crocodile, narwhal horns, skeletons, geodes, ostrich eggs, stalactites, shoes, shells, armillary spheres, telescopes, *Smicroscopia*, fossils, and everything else that caught Kircher's boundless fancy. He installed an intercom at the entrance to the Collegio, a long brass trumpet embedded in the wall and connected to his study. Another built-in megaphone opened out behind a statue in the *Musaeum*; playfully, he called it the Delphic Oracle, and the name stuck.

The Collegio's observatory tower still acts as one of Rome's important weather stations, but with Kircher and his students at work on the premises, the whole building must have bristled with strange appliances. Siphons hung down the walls, telescopes studded the roof, the *Musaeum*'s instruments blared forth their melodies. The Jesuits' rooms were filled with trick mirrors, barometers, thermometers, *Smicroscopia*, clocks, and dancing bronze cherubs encased in glass bulbs. Their word for experiment was "experience," just the right expression for this wholehearted plunge into the mysteries of nature. Not surprisingly, Kircher's students warmed to his enthusiasms with passionate loyalty, and several went on to illustrious careers of their own.

Father Athanasius also gave public lectures. When the Society of Jesus celebrated its centenary with a general convention in 1640, he treated the mathematicians among the delegates to a technical disquisition on Noah's Ark, calibrating the length of the biblical cubit, using Galileo's work on floating bodies to help determine the ark's buoyancy, hypothetically partitioning off the animals' cabins to keep clean beasts from unclean, large from small, and to manage their waste efficiently (it was poured down grates into the ark's bilge to feed the snakes). Surely he must have chosen to illustrate his talk with the help of a magic lantern, a device he is often credited with inventing (he did not, but he used it more effectively than anyone else in his time). Typically the magic lantern in Kircher's *Great Art of Light and Shadow* (*Ars magna lucis et umbrae*) is shown projecting the image of a soul in Purgatory to an audience of simpler folk than the Jesuit mathematicians. It is no accident that in the 1970s the Jesuits were the first to outfit a nativity creche in Rome with laser-generated special effects.

Something of the flamboyance of Kircher's extravaganzas was captured for an international audience—and for future generations—by his remarkable books. These, more than forty in all, range in size from ponderous folios to a tiny volume on magnetism dedicated to a Mexican divine. Most of them are lavishly illustrated, often from his own charmingly awkward sketches. Comparison of these refined, expensive volumes with Kircher's original drafts testifies to the skills of seventeenth-century engravers, who made his tormented efforts at art look supremely elegant. Some of the illustrations, like Lievin Cruyl's images of ancient Babylon for Kircher's *Tower of Babel* (*Turris Babel*); the suns, moons, and maps of his *Subterranean World* (*Mundus subterraneus*); and the panoramic views of Rome's chief obelisks for his *Pamphili Obelisk* (*Obeliscus Pamphilius*) and *The Celebrated Museum* (*Musaeum celeberrimum*), rank among the very best examples of the printer's art.

Kircher's first publication in Rome, *The Coptic, or Egyptian, Forerunner* (*Prodromus Coptus*) of 1636, presented a brief introduction to Coptic, the liturgical language of Egyptian Christians, written in an alphabet adapted from Greek during the latter days of the Roman Empire. Although the Vatican Library had assembled a large collection of Coptic manuscripts over the centuries, almost no one in seventeenth-century Rome was able to read them; with the powerful Ottoman Empire in control of Cairo and most of the eastern Mediterranean, contact between Italy and Egypt had become precarious, dependent on crossing seas patrolled by the clashing forces of the Turkish navy and the Knights of Malta, and marauded by legions of pirates, both Christian and Muslim. In Europe itself, potential new readers of Coptic would have been further frustrated by the absence of any dictionaries or grammar books to help them with a language that bore only scant relation to any others they were likely to know. But Kircher had acquired a medieval Arabic manuscript that provided an introductory grammar, of which the *Coptic Forerunner* offered a partial translation into Latin.

For Kircher, the point of learning Coptic was simple: it descended, he claimed, from ancient Egyptian, and hence held the answer to deciphering the hieroglyphs. His book's full title made both the connection and his own progress with decipherment clear to one and all: *The Coptic, or Egyptian, Forerunner . . . in which Both the Origin, Age, Vicissitude, and Inflection of the Coptic or Egyptian, Once Pharaonic, Language, and the Restoration of Hieroglyphic Literature Through Specimens of Various Paths of Various Disciplines and Difficult Interpretations Are Exhibited According to a New and Unaccustomed Method.*

Appropriately, Kircher dedicated the *Coptic Forerunner* to his sponsor, Cardinal Francesco Barberini. A fellow Jesuit, Melchior Inchofer, acted as censor, praising the work with extravagant enthusiasm. Most contemporary censors' statements simply made laconic declarations that a book contained nothing contrary to faith, but Inchofer, in a striking departure from that businesslike norm, offered effusive expectations for the future of Egyptian studies in Rome, hailing the book as "a worthy beginning from which we may anticipate what will follow."

Thanks to the efforts of Cardinal Barberini, the *Coptic Forerunner* was published by the official press of the papacy's missionary arm, the Holy Congregation for the Propagation of the Faith (Sacra Congregatio de Propaganda Fide), an institution founded in 1622 and strongly promoted by Pope Urban VIII. Both the Collegio and the Society of Jesus devoted intense effort to the search for a universal language by which they might communicate the Gospel to the wider world, and in this search Egyptian hieroglyphs, with their easily recognizable images, had always seemed to present a potential model for a universal script—if only someone could understand what they said. Kircher's "worthy beginning" looked like the outline of a solution, and in a certain sense it was, for he was quite correct in surmising that Coptic descended from ancient Egyptian, and was among the very first scholars to say so.

This same search for a universal language prompted Kircher's simultaneous experimentation with hieroglyphs of his own making, to create an early version of symbolic logic. Meanwhile, other Jesuit fathers explored communication by gesture, with disappointing results among the heathen and revolutionary success among the deaf; American Sign Language, among many others, in fact derives directly from these early Jesuit gestural languages.

Aside from its papal backing, the press of the Propaganda Fide presented another immediate advantage for the *Coptic Forerunner* and its author—it had a set of fonts in exotic types: Greek, Roman, Coptic, Arabic, Hebrew, Rashi Hebrew, Ethiopian. These were cheap and serviceable, and at least allowed Kircher to bolster his credibility as a reader of ancient Egyptian by displaying his command of eleven other languages. He would exploit the same technique for the rest of his career, as he added another twelve languages to his repertory. But he also began an aggressive search for the best typography money could buy: better fonts, better paper, better illustrations, larger formats. In this search he would meet with phenomenal success.

As the title implies, Kircher intended his *Coptic Forerunner* as the prelude, first to a full translation of his Arabic manuscript, and then to a much more comprehensive study of Egyptian antiquity. His prefaces to Cardinal Barberini and to the "Benevolent Reader" reveal that he was already planning to call this great definitive work *The Egyptian Oedipus* (*Oedipus Aegyptiacus*), comparing himself to the ancient Greek hero who solved the riddle of the Theban Sphinx:

> I made this beginning [he told the Cardinal] for no other purpose than in order to demonstrate publicly what I was going to reveal in my subsequent works, and how immense a light the knowledge of foreign languages brings to occult studies and disciplines. For persuade yourself that another work (which shall be as superior to this one as Philosophy is usually regarded as loftier and more sub-

lime than Grammar) on the Hieroglyphic Riddles, I say, a new revived Oedipus, will easily supply every lack.

Before he could make further progress with his projects, however, Kircher received the missionary call that every Jesuit was bound by vow to obey without question. A German princeling of twenty-two, the Landgrave of Hessen, converted to Catholicism in 1637, and the pope, in order to exploit this occasion to the fullest, decided to groom him as a cardinal. Kircher was assigned to accompany the young man as father confessor on an educational trip to Sicily and Malta; the illustrious German scholar (and future Vatican librarian) Lucas Holstenius came along as tutor.

Kircher's trip with the Landgrave would be his only extended journey after arriving in Rome. He desperately wanted to go as a missionary to China, a desire the order scrupulously ignored. By deliberate plan, the society kept its most brilliant minds close to home in the belief that these fathers were the most likely to poke their heads into trouble, throw themselves into local customs, and forget their Ignatian training in an access of intellectual enthusiasm. Furthermore, even by the standards of his tumultuous age, Kircher seems to have had an uncanny propensity toward disaster. He craved the violence of nature, and nature obliged him with rare consistency. As he discovered on his trip with the Landgrave, Southern Italy and Sicily especially were filled with natural wonders: as their party passed through the Straits of Messina, the treacherous tides of Homer's Charybdis snatched at their Maltese galley. Mount Etna erupted as they pulled into Catania, where Kircher eagerly climbed down the smoking crater and made a lurid sketch of its red-hot lava pouring forth. The island of Stromboli and Vesuvius performed nearly as well on the way home; an earthquake leveled parts of Calabria. By the time his trip ended in 1638, Kircher was obsessed with volcanoes.

Malta, where the Landgrave and his entourage spent a few months, presented its own peculiar microcosm. Its stark limestone outcrops had served as home to the military order known as the Knights Hospitallers of Saint John since the 1530s, after their expulsion from Rhodes by the Ottoman Turks. Arid and isolated, Malta nevertheless boasted two signal strategic advantages as a naval base: its position in the very center of the Mediterranean, and its incomparable harbor of interconnected bays. From this well-protected base, the knights picked off Turkish pirates and rattled sabers at the Ottoman navy for centuries. After a brutal Turkish siege in 1565, they had built the Maltese bastions into an unassailable fortress that served effectively to maintain a balance of sea power between Islam and Christendom. The Inquisition had also established an office in Malta, where in 1637 a Sienese prelate named Fabio Chigi presided as Apostolic Delegate and Chief Inquisitor. It was an important post for a highly promising prelate of just under forty. Yet if Chigi's service among this peculiar community of aristocratic Christian warriors had great symbolic, and

> *... [E]ven by the standards of his tumultuous age, Kircher seems to have had an uncanny propensity toward disaster. He craved the violence of nature, and nature obliged him with rare consistency.*

probably strategic, importance, he himself regarded his term as little better than exile. A scholar, an amateur architect, and a consummate diplomat, he confessed in his letters to missing the excitement of Rome. The entry of Kircher and Holstenius into his lonely outpost resulted in the forging of friendships that would greatly affect the lives of all three when, eighteen years later, Fabio Chigi became Pope Alexander VII.

From Kircher's standpoint, Malta also presented yet another new world of curiosities: its language, based on Phoenician and littered with Italian loan words; its geological oddities; and the ostrich farms that dotted its arid landscape. For the knights, he designed a new machine, the Maltese Observatory, to whose intricate workings his student Gaspar Schott would dedicate a chapter of his own book on modern technology, *Technical Curiosities* (*Technica curiosa*). Unfortunately, neither Schott nor Kircher ever recorded an illustration of this marvelous contrivance, but reportedly, among other useful attributes, the Maltese Observatory traced the progress of the sun, moon, and planets; tallied the dates of movable feasts on the Christian calendar; told time throughout the world; and determined auspicious times to go fishing, take medicine, or give birth. Inscribed in twelve languages (Hebrew, Chaldaean, Syrian, Arabic, Ethopian, Coptic, Greek, Latin, Italian, French, Spanish, and German), it also combined three ideal geometric forms: circle, cube, and pyramid, as its proud inventor reported in *The Maltese Observatory* (*De specula Malitensi*):

> The first part is circular, a base, as it were. It displays the horizon, information about the winds, and the practice of the art of navigation. . . . The central part, or cube, with its five sides, that is, four vertical faces and one horizontal parallel to the horizon, contains, by a singular method, whatever is pertinent to the Mathematical and Physio-logical [*sic*] Sciences. . . . The pyramid has four faces corresponding to the cardinal directions. . . .
>
> Receive it, Generous Knights, and receive it with favorable minds and eyes!

Just what Athanasius Kircher and Fabio Chigi discussed in the course of their encounters on Malta never appears in the correspondence of either; they were both too professionally discreet to have confided such conversations to their archives. Nevertheless, subsequent events would attest to the lasting friendship they made on that curious island. They were virtual contemporaries (Chigi was four years older). Their similarities of interest ran deep. To a profound degree, Chigi shared Kircher's awareness of spectacle as a means of persuasion; he had also begun to recognize, and to exploit, the power of the press, and may have been instrumental later in helping Kircher to do the same. His connections with the Society of Jesus had always been close and would remain so. A man of penetrating curiosity and cruelly precarious health (he was tormented for most of his life by kidney stones), Chigi learned to harness Kircher's physical energy in order to further his own intellectual pursuits; he was perhaps the only person who ever taxed his German friend's boundless verve to the point of exhaustion.

Toward the end of Urban VIII's papacy, Kircher brought out a pair of books from the Roman press of Ludovico Grignani. Although

DETAIL, CAT. 79

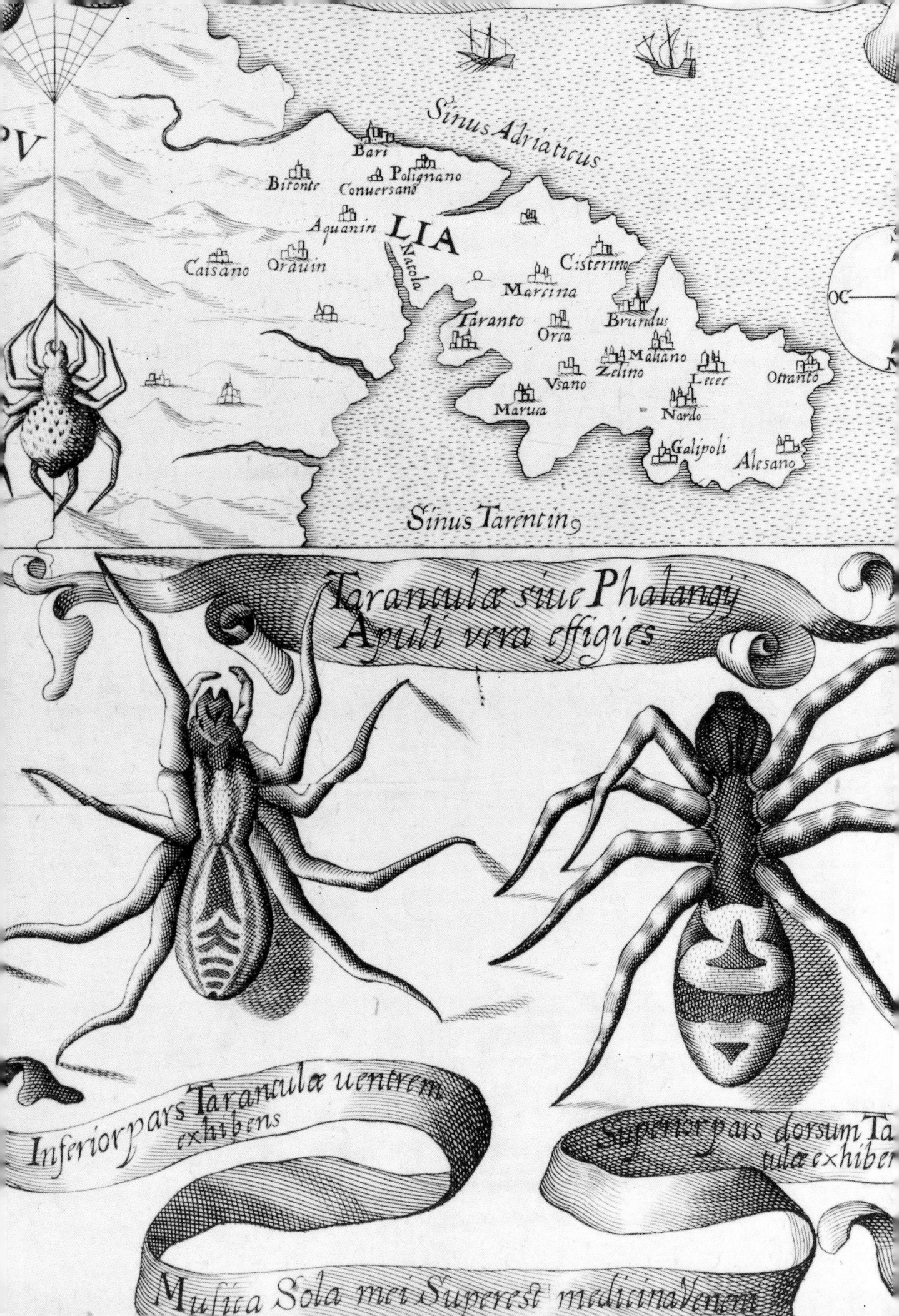

they were issued two years apart, in 1641 and 1643, he claimed explicitly to see them as companion pieces, reflecting two sides of his activity in the Eternal City. At the same time, these stout quarto volumes, one with exotic typefaces and the other with copious illustrations, represented an expensive new level of publication, financed by a new sponsor: Emperor Ferdinand III of Austria, who had begun to take a serious interest in Kircher's investigations as early as 1640. It was the beginning of a long, fruitful partnership. Although Kircher had not been sent to the Viennese court in 1634 and never visited in the subsequent decades, he managed to assert his intellectual presence from a distance through his voluminous writings, and the Hapsburg court was clearly pleased to foster this curious protege in Rome.

The first book of this sumptuous pair, *The Magnet, or the Magnetic Art* (*Magnes, sive de arte magnetica*) of 1641, provides a revealing glimpse of life in the experimental cauldron of the Collegio Romano, as well as nostalgic memories of Kircher's travels with the Landgrave through the southern reaches of Italy. The Jesuits in the 1640s continued to strive valiantly as natural philosophers in competition with the rest of Europe, and *The Magnet* addresses many of the same issues as Galileo's contemporary study of mechanics, *Two New Sciences* (*Discorsi, e dimostrationi matematiche: intorno à due nuove scienze*). The perceived importance of Kircher's essay on magnetism may be gauged by the fact that it was reprinted in Germany only two years later, and again in Rome in 1654. Magnetic attraction played the role in Kircher's world system that gravity was beginning to assume for Galileo and would definitively hold for Newton; it was the force that pulled the planets into their orbits and kept them in motion. He portrayed this mutual pull of physical bodies in vivid terms as the manifestation of God's love: "uniting and connecting everything among us and in the universe."

Magnetism also served to explain phenomena that modern scientists would regard as quite unrelated, such as the motion of plants—a sunflower's tendency to face the sun, for example, or ivy's tendency to twine. Magnetism created the communal and sexual life of animals. The powers of music were, at their heart, magnetic powers. Through musical magnetism, Kircher explained, the strains of the tarantella could draw the toxins of a tarantula's bite from the human body, and *The Magnet* helpfully supplied the efficacious tune in a lovely engraving replete with spider, Sicilian landscape, and musical text.

In Kircher's second large book, *The Egyptian Language Restored* (*Lingua Aegyptiaca restituta*) of 1643, he presented a complete translation of his Coptic-Arabic manuscript; he took the opportunity as well to include extensive revisions to the *Coptic Forerunner*, published seven years earlier. Now at last, as he declared in his preface to what was in effect a textbook of the Coptic language, he could embark upon his comprehensive Egyptological treatise, *Egyptian Oedipus*, in earnest.

> *Magnetic attraction played the role in Kircher's world system that gravity was beginning to assume for Galileo and would definitively hold for Newton; it was the force that pulled the planets into their orbits and kept them in motion.*

*The Egyptian Language Restored* was the last work that Kircher wrote under Urban VIII, who died in 1644, beleaguered and bitter from the repercussions of the Galileo affair, the Thirty Years' War, and a disastrous attempt at home to seize feudal properties by force of arms from some of Rome's most powerful baronial families. With their papal protector dead, the entire Barberini family quickly slipped into exile, Cardinal Francesco among the refugees. But Kircher had already learned the Roman courtier's art of switching allegiance to the reigning pope. By the time Cardinal Barberini rushed off to Paris, the Collegio Romano's resourceful polymath had already made other important alliances, including one with the newly elected Pope Innocent X.

In many ways, Pope Innocent X Pamphili was Urban's direct opposite. Politically, he favored Spain rather than France. Personally, he impressed his contemporaries as a gruff pragmatist rather than an urbane courtier, all the more so because of his comparatively advanced age. Diplomatically, his hard-headed competence represented a welcome change from Barberini's showy ineffectuality; above all, through his nuncio Fabio Chigi, he helped to push through the Peace of Westphalia (1648) that put an end to the Thirty Years' War. At the same time, however, Innocent maintained the Barberini papacy's preoccupation with the arts and spectacle as a means to promote the image of Catholic Rome, especially as the Jubilee Year of 1650 approached in a Europe newly at peace. Under the vigilant eye of his piercingly intelligent, domineering sister-in-law Donna Olimpia Maidalchini, Innocent imposed his own distinct and discerning taste as a patron of the arts. When his attention fell upon Athanasius Kircher, Kircher was well prepared to meet the challenge.

Kircher's *Great Art of Light and Shadow* (*Ars magna lucis et umbrae*), printed in 1646, must have been in preparation long before Pope Innocent ascended his throne; again Kircher used the Roman press of Ludovico Grignani and Grignani's associate Hermann Scheus, but they now moved into a sumptuous folio format with still more copious illustrations than the smaller, but extensively illustrated, *Magnet*. In this new work, Kircher laid claim to inventing the magic lantern, explored the cause for fireflies' glow, and traced the origin of brightness to God, defining light as "the exuberance of God's great goodness and truth."

His ultimate source for this way of thinking was Plato, as modified through Christian theology, and in true Neoplatonic fashion he acknowledged that two of the five senses induced humanity to know God: sight and hearing. He published his first work on the latter faculty, *Universal Music-Making* (*Musurgia universalis*), in 1650, the Jubilee Year. By this time, Kircher, fortified by the Austrian crown's reliable subsidies, had mastered the art of the picture book. His fantastic images of the machines he contrived for the Collegio Romano mingled with allegorical devices, like his rendering for the *Universal Music-Making* of the seven days of Creation conceived as a mighty cosmic pipe organ. On the printed page as in the magic-lantern-lit halls of the Collegio Romano, the music of the spheres could boom out its melodies, and Kircher provided the musical notation for all who could read it:

> I recognized that Music was nothing other than, as Plato says, "knowing the order of all things," and also contemplated this world of the senses as nothing other than a brilliant mystic ten-note chord . . . by which the All-Wise Creator, using a mixture of Consonance and Dissonance, that is, harmonic proportion, produced that wondrous harmony and concord of all Nature that all the Philosophers up to this time cannot admire enough.

In all these endeavors, Kircher continued to pursue the overarching goals that had dominated his intellectual life from the outset: a good Jesuit, he strove to embody Christian doctrine by carrying all its varieties of knowledge within his own trained and capacious memory; at the same time, he sought tirelessly for the universal language by which he might best convey that doctrine to the world at large. A good Christian, he took the world's variety as proof of God's greatness and of Scripture's enduring validity. A good Catholic, he sought to classify that variety under broad, inclusive categories. A good scientist, he applied his theoretical learning to the evidence of his senses. Inevitably in seventeenth-century Rome, these impulses, all powerfully expressed in Kircher's own nature, all urgently fostered by the different communities among which he moved, came into impossible conflict. For a man of Kircher's imagination, the opportunities for trouble with the most inflexible forces in his Church were nearly as rife as they had been for Galileo in the decades before. But rather than engage in direct controversy, Kircher began an operation of a very different order. His books began to play word against image, saying one thing in their texts and another in their pictures. He claimed in print to agree with Plato that nothing in the phenomenal world was real, and yet at the same time his instrumentation and his experiments focused on that phenomenal world with relentless precision. In a sense, Kircher employed the same dodge by which Galileo and his fellow Copernicans had been able to discuss the position of the sun in the universe: by treating their speculations as hypothesis rather than truth. Galileo had finally tired of this continuing dissimulation and paid the price in enforced silence. Kircher, by contrast, moved beyond hypothesis into the realm of outright fiction.

*The Magnet*, the *Great Art of Light and Shadow*, and *Universal Music-Making* all show how Kircher sought to explain the evidence of his senses through hypothesis—the method of natural philosophy—while entrusting any general explanations to the symbolic images of allegory—the method of theology and literature. By claiming to use symbols and parables, he could tell fictional tales about verities he could not prove, whether because the limits of his instrumentation prevented him, or because the dictates of his order commanded him to hold his peace about sensitive topics. He swathed his scientific speculations in such striking showmanship that the sources for his work seemed obvious, parading his learned citations in dozens of languages, his exotic type fonts, his lavish engravings, his machines, his apparent command of every bit of information worth knowing. And yet, just as he had always done in his youth, Athanasius Kircher kept a stoical silence about what lay closest to his heart. Silent he may have been on many accounts, but Kircher was also by vocation a communicator.

He locked away his forbidden readings and his radical thoughts within his performances, but he also provided the key to what he was about and placed it among the hieroglyphs of ancient Egypt.

Before the press of Vitale Mascardi could produce Kircher's monumental four-volume *Egyptian Oedipus*, the tireless father, now happily ensconced in a rhythm of phenomenal productivity, issued yet another preliminary study on Egyptian antiquity, this one dedicated to the pope himself on the occasion of the 1650 Jubilee. The *Pamphili Obelisk* (*Obeliscus Pamphilius*), a folio tour de force from the Grignani press, commemorated the erection of an Egyptian obelisk in front of the Pamphili pope's family palazzo on Piazza Navona. The obelisk itself had begun its Roman career as an ornament in the sanctuary of Isis—brought from Egypt by Emperor Domitian (reigned 82–96 C.E.). In the early fourth century, just before losing to Constantine at the Battle of the Milvian Bridge (311 C.E.), another emperor, Maxentius, had transferred the granite needle to his own circus along the Appian Way, where it lay in pieces on the ground until Pope Innocent decided to move it yet again. Piazza Navona was an appropriate setting; the obelisk's long, round-headed outline of this expansive urban space marked the site of the Circus of Domitian.

Here in 1648, the sculptor Gianlorenzo Bernini set the restored obelisk atop a delightful Fountain of the Four Rivers that gave full rein to his genius as designer and technician alike. Kircher, for his part, translated its hieroglyphic inscriptions and summarized them for engraving on the four granite plaques that still decorate the obelisk's base. He probably did a great deal more, for it was his interpretation of the ancient Egyptian texts that guided Bernini's design for the fountain, and the hollow mountain from which the fountain's four rivers gush follows Kircher's ideas about the structure of continents—he believed that all mountains stood domelike above huge underground reservoirs (he called them *hydrophylacia*) that fed the rivers of the world.

The *Pamphili Obelisk*, despite its superb engravings of fountain and pope, was not primarily designed as an occasional pamphlet, however much it resembled one. Instead it offered a prelude to the imminent *Egyptian Oedipus*; Kircher was by now an astute publicist, for his order and for his own work. More significantly, however, this 300-page study offered clues by which a world-weary, no-nonsense pope might come to understand the meaning of all of Kircher's labors to date, from his natural-philosophical treatises to his Egyptological texts.

*The* Pamphili Obelisk (Obeliscus Pamphilius), *a folio tour de force from the Grignani press, commemorated the erection of an Egyptian obelisk in front of the Pamphili pope's family palazzo on Piazza Navona.*

For the first time, a little Egyptian figure made his appearance among the Jesuit's images, never to disappear again: Harpocrates, the infant god who raises his finger to his lips as an injunction to silence. True wisdom, Kircher insisted to the pope, shunned expression of the naked truth; Pythagoras, the first Greek philosopher, had learned that much from the ancient

Egyptians and passed the insight into Western tradition:

> The Pythagoreans conveyed the teachings that their Master had learned from the Egyptians through riddles and symbols, reckoning that naked and open exposition was inimical to God and Nature . . . and they persuaded themselves and firmly believed that God withdrew himself from the senses of common, profane humanity, hiding understanding and knowledge beneath likenesses and parables of various sorts. On the other hand, it would be welcome and acceptable to Him that those genuinely desirous of true wisdom should investigate his hidden mysteries along secret paths, and proceed to uncover the secret sacraments of His holy doctrine by this underground way.

*. . . a little Egyptian figure made his appearance among the Jesuit's images, never to disappear again: Harpocrates, the infant god who raises his finger to his lips as an injunction to silence.*

Ostensibly, Kircher's remarks applied Pythagorean principle to reading the hieroglyphs of the Pamphili Obelisk. But he adapted this same form of interpretation to the reading of Scripture:

> The Rabbis say that all of Holy Scripture is nothing other than an extended symbol of the most sublime matters and mysteries, appropriate only to learned men long and deeply versed in the Law so that they know it. So, too, Christ our Savior conveyed this same eternal Wisdom in the form of speech [known as] parable, as we often read among the Gospel writers. Thus the hidden substance of God does not know how to enter profane and polluted ears by means of naked speech. Julian the Apostate, although impious, rightly said that "Divine Nature loves to be covered and hidden away."

Galileo's problems with the Inquisition had formally hinged on the interpretation of Scripture, on whether its words were literally or figuratively true, and specifically whether the text of the Bible could be made to conform with a stationary sun at the center of the cosmos. The Tuscan astronomer had argued forcefully on several occasions that Scripture was designed for moral guidance, not to render a factual account of the structure of the universe. It was an argument that he in turn had taken from the Southern Italian philosopher Giordano Bruno, who had been burned at the stake in Rome in 1600 for heresies far more radical than Galileo's.

Galileo was silenced for declaring that the earth moved around the sun, and for daring to show how Scripture could be made to confirm the fact; Bruno, on the other hand, identified the Copernican solar system as but one among a numberless multitude of orbital systems, all whirling within an infinite universe composed of atoms in constant flux. As for Scripture, it had nothing to do with literal phenomena at all—it offered guidelines for conduct, not an account of nature. Although the Inquisition banned Bruno's writings when it sentenced him to death, scholars such as Kepler in Northern Europe, at least, seem to have had continuing access to his work, and they read it with a combination of interest and philosophical terror—Bruno's cosmos was far too large and too strange for his generation and for many generations after his. But even in Italy, the legacy of the burned philosopher persisted despite the Inquisition's best efforts to suppress it. Galileo, as Kepler noted in 1610, did read Bruno thoroughly and borrowed some of his ideas. So, evidently, did Athanasius Kircher.

Bruno anticipated Kircher's statements in the *Pamphili Obelisk* about the hidden allegories of Nature and Scripture, and had likewise emphasized the debt of Christian scriptural wisdom to Egypt, Pythagoras, Plato, and the rabbis of the Talmudic tradition. Bruno drew his formulations from extensive readings of ancient figures like Plato, the medieval mystic Ramon Llull, and Renaissance philosophers such as Marsilio Ficino and Nicholas of Cusa; Kircher's works carefully cite these same sources without, however, emphasizing the mediating figure to whom his own written work owed something of its scale and vividness. Yet by a small detail like citing the renegade pagan emperor Julian the Apostate, Kircher served notice to attentive readers, among whom he presumably included Pope Innocent (and the latter's recently appointed secretary of state, Fabio Chigi), that even doctrinally wrong authors could speak the truth. In effect, therefore, Kircher's instructions for reading the secret lore of the hieroglyphs marked out the "subterranean way" toward reading Kircher himself.

From 1652 to 1654, the four stout folio volumes of *Egyptian Oedipus* finally appeared, "a Work," the author told his "Benevolent Reader," "born of twenty years of a continuous mental firestorm." *Oedipus*, both for its content and for its unprecedented scale, marked yet another stage in Kircher's development as a publishing performer. Ironically, the *Pamphili Obelisk*, designed to look as if it belonged to the slight, bombastic ranks of occasional literature, contained some exceedingly sober scholarly analysis in addition to its messages about interpretation. *Oedipus*, on the other hand, disguised as a definitive encyclopedia of ancient Egypt, put on a monumental show. Its engraved frontispiece shows Oedipus as a classical Greek monarch, a swaggering version of Alexander the Great, with his flowing hair and scanty chiton, utterly remote from Sophocles' toweringly tragic Oedipus Rex. This brash young hero, who has dispatched the riddle of the Sphinx as swiftly as Alexander sliced open the Gordian Knot, stands quite obviously as an alter ego of black-robed, bearded, fiftyish Athanasius Kircher, S.J. The diffident German novice who once seemed so stupid is, as he might have said, "parasangs away."

The *Oedipus* was the work on which Kircher's stature as a reader of hieroglyphs would stand or fall in the world at large, and he knew it. He began, therefore, with a seventy-page collection of testimonials from all over the globe, agreeing in a babel of languages and scripts that at last the wisdom of Egypt lay accessible to all inquiring minds. But Kircher's hieroglyphic breakthroughs were, unfortunately, largely illusory. His sources for the hieroglyphs themselves were hopelessly inaccurate: most importantly, they included a treatise on the script attributed to one Horapollo, a text composed in Egypt at the very end of the Roman Empire (fourth century C.E.). This tract, whose rediscovery had caused a considerable stir in the fifteenth century, was written in Greek, the bureaucratic language first imposed on Egypt by the Ptolemies and continued by the Romans. The long ascendancy of Greek in Egypt meant that scholars like Horapollo knew only the barest remnants of hieroglyphics. A more impressive work of ancient scholarship, Plutarch's essay *On Isis and Osiris* (written in the second century C.E.), supplied the meaning for a handful of hieroglyphic symbols, but Plutarch

# ORTHOGRAPHIA
## PRÆCIPVÆ DOMVS ARCIS VRANIBVRGI IN INSVLA PORTHMI DANICI VENVSIA Vulgo HVENNA, ASTRONOMIÆ INSTAVRANDÆ GRATIA CIRCA ANNVM 1580, à TYCHONE BRA- HE EXÆDIFICATÆ.

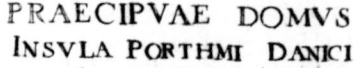
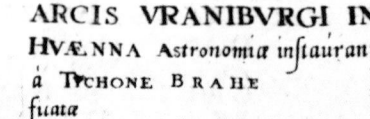

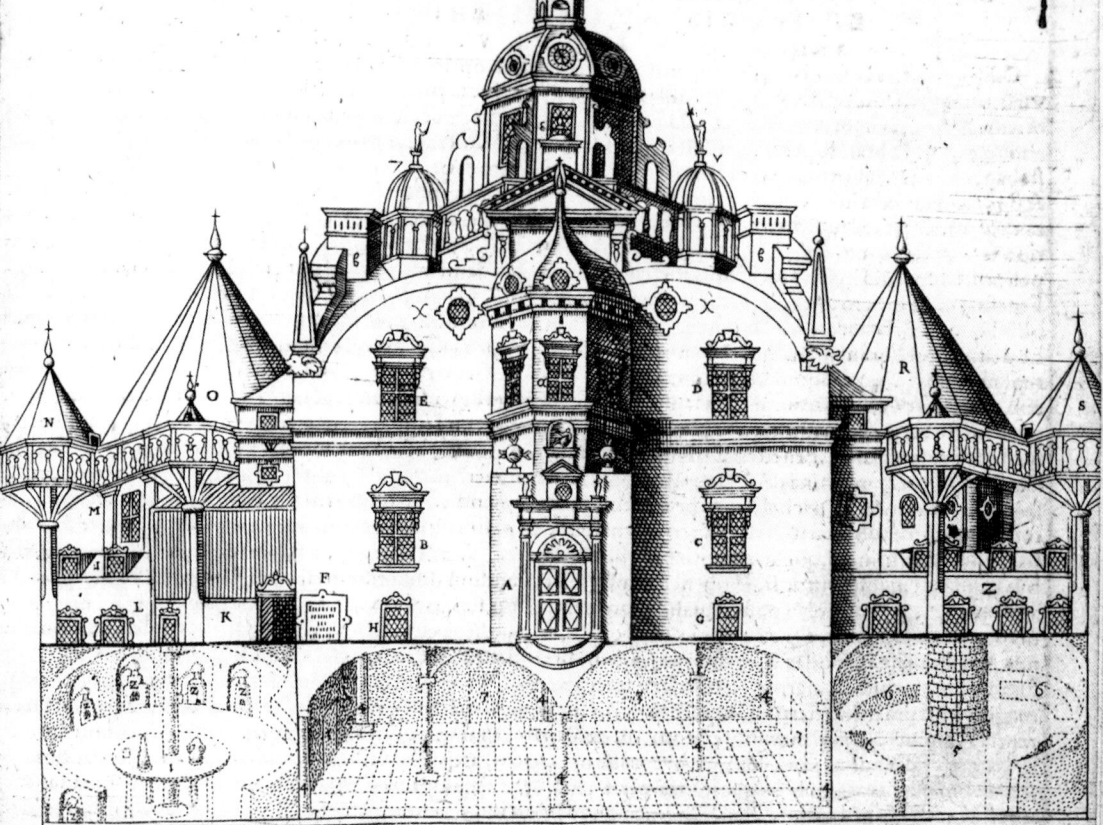

## ICHNOGRAPHIA ET EIVS EXPLICATIO

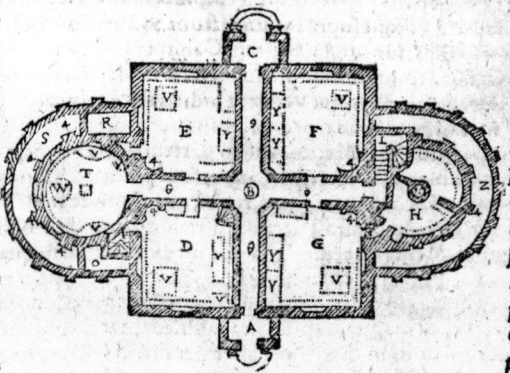

A Ianua Orientalis. C. Occidentalis. ☉. Trãsitus 4. ad angulos rectos cõcurrentes, ut Cænaculũ hybernũ siue hypocaustum D. ampliarétur, idq; in eius angulo post fornacẽ paruũ quoddam et secretum laboratorium spagyricum esset, in quo tñ quinq; distinctim erant furni, q promptius ad manus isthîc operi Pyronomico inseruiebãt, ne sem per in maius illud descendendũ foret B. Fons æquarium volubi lem rotans, qui aquas hinc inde cum lubuit, in sublime eiaculaba tur. D. Cœnaculum illud hyber num. E. F. G. Cameræ pro hospi tibus. L. Gradus pro ascensu in superiorem contignationem. H Coquina. K. Puteus comentitius 40. ulnas profundus, artificio hydraulico seruiens et aquas per siphones hinc inde occultè per murũ transeũtes in singulas Cameras tam superiores quam inferiores distribues P. Gradus pro descensu in laboratoriũ Chy micum T. Bibliotheca. VV. Globus magnus Orchalcius nuper exhibitus V. Quatuor Mensæ pro Studiosis, 4. Camini tam è laboratorio inferiori ascendentes, quam in quatuor angulis conclauium Y. Lecti in iisdem conclauibus. hinc inde dispositi. Cætera actus inspector propria intentione facilè discernet. Intelligenda autem sunt hæc omnia in ea quantitate, veluti fundamento maioris domus supra depictæ quadrare poterunt: Licet hic coarctationis loci gratiã in duplo quasi minori forma exhibeantur.

made no pretense to knowing Egyptian. Kircher also made extensive use of an archaeological artifact called the Mensa Isiaca—the Table of Isis, a bronze tabletop inlaid with silver Egyptian designs that had been excavated in Rome on the site of the ancient temple of Isis in the 1520s and purchased by the Venetian writer (and future cardinal) Pietro Bembo. Unfortunately, it is now clear that the tabletop must have been made by a Roman craftsman for a Roman devotee of Isis, for the hieroglyphs that Kircher and his contemporaries studied with such close attention are sheer decorative fabrication: an ancient Roman's idea of what Egyptian writing might look like. Nor was Kircher's command of Coptic sufficient on its own to penetrate the Egyptian script; as we know, it would take a linguist of gifts comparable to his—Jean-Louis Champollion—together with the Rosetta Stone, with its literal translations from Greek to Egyptian, to plumb the final riddle of the Egyptian Sphinx.

However, Kircher's readings of the hieroglyphs did make the Egyptians say all the right things; they had boasted of their superior wisdom and their religious insight to every denizen of the Mediterranean from the time of Homer (who made the Old Man of the Sea an Egyptian) to the ages of Plato, Caesar, Saint Paul, and Plutarch. As Kircher deciphered their sacred script, the people of the Nile continued to prove their mettle as repositories of primeval religious lore, dispensing wise sayings and civilized advice, although he could not help adding, in good Judeo-Christian vein, that their worship of animals was reprehensible: "I am truly amazed that it was possible for people otherwise of sound mind ever to have accepted, let alone approved, such insane and fanatical hallucinations."

Not everyone was convinced by Kircher's readings, however. That strange, bitter Jesuit, Melchior Inchofer, who as censor of Kircher's *Coptic Forerunner* had expressed such enthusiasm for the work, seems to have soured on his colleague's enterprise some years later. Inchofer's 1645 satire of the Jesuit Order, *The Monarchy of the Solipsists*, already presented a figure whose resemblance to the great performer of the Collegio Romano seems unmistakable, an "Egyptian wanderer," who, seated on a wooden crocodile, "broadcast trifles about the Moon."

Elsewhere in the same diatribe, Inchofer described the researches pursued by Solipsist, that is, Jesuit, natural philosophers. Although he died in 1648, before the publication of either the *Pamphili Obelisk* or the *Oedipus*, it is hard not to associate the beginning of the following passage from Inchofer's satire with an image that Kircher used in both books: a scarab rolling its ball of dung through the planetary spheres:

> "Philosophical works among [the Solipsists] are more or less of this sort: "Does the scarab roll dung into a ball paradigmatically?" "If a mouse urinates in the sea, is there a risk of shipwreck?" "Are mathematical points receptacles for spirits?" "Is a belch an exhalation of the soul?" "Does the barking of a dog make the moon spotted?" and many other arguments of this kind, which are stated and discussed with equal contentiousness. Their Theological works are: "Whether navigation can be established in imaginary space." "Whether the intelligence known as *Burach* has the power to digest iron." "Whether the souls of the Gods have color." "Whether the excretions of Demons are protective to

CAT. 57

humans in the eighth degree." "Whether drums covered with the hide of an ass delight the intellect."

Inchofer knew his target well; he had acted as censor for Kircher's *Great Art of Light and Shadow*, which devoted a chapter to the color of angels (described by some Neoplatonic writers as "the souls of the Gods"). *The Magnet* and *Universal Music-Making* discussed the effects of music, including the music produced by drums covered with the hide of an ass, on the intellect. As for navigation into imaginary space, it would form the pretext for Kircher's next book.

Cosmology, by tradition, had been a chief concern of the Collegio Romano's professors of mathematics: Clavius, Grienberger, Grassi, and Scheiner prominent among them. Like his predecessors, Kircher followed the telescope's revelations with avid interest, but as the first professor to serve after Galileo's condemnation, he reacted more strongly to the constraints the Church had imposed upon such curiosity. Because of its very subject matter, light, Kircher's *Great Art of Light and Shadow* had already treated the sun, moon, and stars—and had done so under the vigilant eye of his censor, Melchior Inchofer, the most vitriolic of Galileo's examiners in the trial of 1633. Inchofer had let the book pass, despite Kircher's insistence that the sun was not the perfect crystalline sphere envisioned by the ancients, but an actively erupting ball of fire. Evidently Christoph Scheiner's hard-fought battle with Galileo over sunspots had at least won the order the right to honor the observations of their own eyes, so that Kircher was permitted to describe the sun as he saw it, "a fiery, rough, and uneven body, [as] has been noted recently by . . . experiment."

To consider the structure of the cosmos, however, was to invite trouble, and Kircher knew it. He had been putting his mind instead to a book on geology, tentatively called *The Subterranean World* (*Mundus subterraneus*), a work inspired by his experience with volcanoes in which he could finally give voice to his delight in nature's curious vagaries and his continuing attraction to its more violent extremes. No longer able to clamber down smoking craters, he contented himself with standing under crashing waterfalls in the Roman countryside and waiting for lightning to strike. In the 1650s, however, he had to face intractable pressure of another sort, which finally drove him to write on the universe after all: the relentless stimulus of the former student who had accompanied him from Germany to Avignon, and thence to the Collegio Romano, Gaspar Schott. As Schott told the story, Kircher had hesitated to write about cosmology because of the novelty of his ideas and the risks they posed, and also because he had been engaged in two other large projects: the *Oedipus* and the as yet unpublished *Subterranean World*. One evening, after his usual late afternoon colloquy with Schott in one of the Collegio's gardens, he attended a concert by three lutenists, whose music lifted him, as it often did, into a trancelike rapture. That night, as he stated himself and reported to Schott, he had a remarkable dream, in which his guardian angel appeared to him and escorted him through the heavens. Prophetic dreams, like Kircher's trances, had been part of his life from an early age, and he had learned to act upon them without question after one such dream predicted the invasion of his Jesuit house in Würzburg by Swedish troops.

A book called *The Ecstatic Heavenly Journey* (*Itinerarium exstaticum coeleste*), published by Vitale Mascardi in 1656, finally brought Kircher's cosmology to light. It was comparatively modest in format, and explicitly touted as a dream vision, both maneuvers designed to diminish any apparent claims to authority. At the same time, however, readers must inevitably have recalled one of the most famous of all dream visions, Cicero's *Dream of Scipio* (*Somnium Scipionis*), in which, like Kircher, the protagonist ascends into the heavens to discover that the souls of all good people dwell among the stars. In his *Ecstatic Heavenly Journey*, therefore, Kircher may have pretended to be writing a forerunner of science fiction, but he did so in a way that also claimed something of Cicero's authority.

Kircher's guide to the heavens is an angel named Cosmiel, a tall youth with shimmering wings and eyes like burning coals, who lifts him beyond the earth's atmosphere and drops him down on the heaving surface of a tempest-tossed lunar sea, composed of the same four elements as the seas on earth. Orbit by orbit, man and angel traverse the planets, finally arriving at the outer limits of the universe in the sphere of the fixed stars. Along the way, Cosmiel delivers himself of some highly unorthodox opinions: to his charge, who was compelled by Jesuit mandate to teach an Aristotelian cosmos, he observes that Aristotle had not told the whole truth about the stars and planets.

Furthermore, Cosmiel's itinerary takes the pair through a sphere of fixed stars that is surprisingly unfixed and unspherical: infinitely deep, it seems to contain pockets, packed with more stars than the Milky Way, whose glittering congestion had been revealed once and for all by the telescope. As for the structure of this universe, Cosmiel and Father Kircher insisted that it conformed to the compromise system devised by the Danish astronomer Tycho Brahe to accommodate both Copernicus and established belief: the planets moved around the sun, while the sun and the fixed stars moved around the earth.

But in fact, Cosmiel escorts a quaking Kircher to a gigantic, pocketed orb of fixed stars that revolves around the sun. Jesuit and angel stand one step away from a Copernican world, as close as they could come without admitting to one explicitly. At the same time, however, their heaven's infinite size and uncertain margins also evokes a still more dangerous universe: the boundless expanses envisioned by Giordano Bruno, that abstract infinitude of space which a fearful Kepler had compared to permanent exile.

*Like his predecessors, Kircher followed the telescope's revelations with avid interest, but as the first professor to serve after Galileo's condemnation, he reacted more strongly to the constraints the Church had imposed upon such curiosity.*

And Kircher's fellow Jesuits noticed the oddity of Cosmiel's cosmos. In 1656, a pair of censors from the order criticized the doctrinal deviations of the *Ecstatic Heavenly Journey* to pointed effect. Father Athanasius may have rejected Copernicus, they complained, but it was a pallid performance at best: "he is evidently prevaricating on the matter," they concluded, "and he is not doing so from the heart, but in order not to say anything openly contrary to the decrees and institutions of the Holy

Roman Church. But he would have been more obedient if he were to deny his opinions, which he does in passing and in word only, beyond every probable reason. . . ." The censors also caught echoes of an infinite universe, although they refrained from tracing it explicitly to Bruno's forbidden texts.

Unlike Galileo, however, Kircher suffered few repercussions from this attack, for unlike Galileo, he managed to stay in the good graces of the reigning pope. Since 1655, that pope had been Alexander VII, the former Fabio Chigi, whose interest in Kircher's work had only intensified since their days together on Malta. On another front, the indefatigable Gaspar Schott flew to his teacher's defense, producing a new edition of Kircher's *Ecstatic Heavenly Journey*, buffered by Schott's own copious apparatus of learned notes and a direct rebuttal to the Roman censors' criticisms. Retitled *Iter exstaticum*, this new edition was published in Germany in 1660 and reprinted in 1671. The German press operated under much more liberal conditions than printing in Rome; among the sources Schott cited outright in support of Kircher's contentions is Giordano Bruno.

This revised edition of the *Ecstatic Heavenly Journey* also features a charming frontispiece, with a beaming Kircher and a solicitous Cosmiel standing off to the side of the universe, whose earth, in the barest concession to Tycho Brahe, hangs like a necklace around a sun that is obviously the center of the whole system.

In 1656, as Kircher faced the censors, Rome and the pope had more urgent concerns than the intricacies of cosmic structure. Plague had broken out in 1655, and Kircher, for the time being, replaced his telescope with the *Smicroscopium*, in eager pursuit of the culprit. He published his results in 1658 in a short study called *Physico-Medical Examination* (*Scrutinium physico-medicum*), one of the world's earliest attempts to trace the disease to microbial contagion. The wriggling microcosm revealed by Kircher's *Smicroscopium* had convinced him that most physical substances were shot through with tiny worms. Plague, he surmised, should be traced to the same kind of agent, "worms . . . so tiny, so slender and subtle" that his *Smicroscopium* could not quite isolate them. Despite the gruesome effects that plague exerted on the population of Rome, Kircher, newly returned from his figurative *Ecstatic Heavenly Journey*, marveled at the scope of Creation, from large to small.

He would summarize all these experiences in his next great book, the long-advertised *Subterranean World* of 1655. Dedicated to Pope Alexander VII, its very format bore witness to the influence of its dedicatee, a man who had labored for years to bring Protestant and Catholic together, who had exchanged the oddly cosmopolitan backwater of Malta for the bustling courts and international diplomacy of Westphalia, and in the process had forged a network of learned connections throughout Europe. For years Fabio Chigi had facilitated his friend's access to Protestant presses. As Pope Alexander VII, he continued to do so, binding the world of learning together despite its differences of creed. The *Subterranean World*, therefore, was published in Protestant Amsterdam by the Dutch firm of Jansson and Weyerstraet, the first in a long series of volumes that Kircher would produce for that active house over the

CAT. 41

## MUNDI SUBTERRANEI

Sect. I.

Pars altera Tabulæ
*Piscium figuras exprimens.*

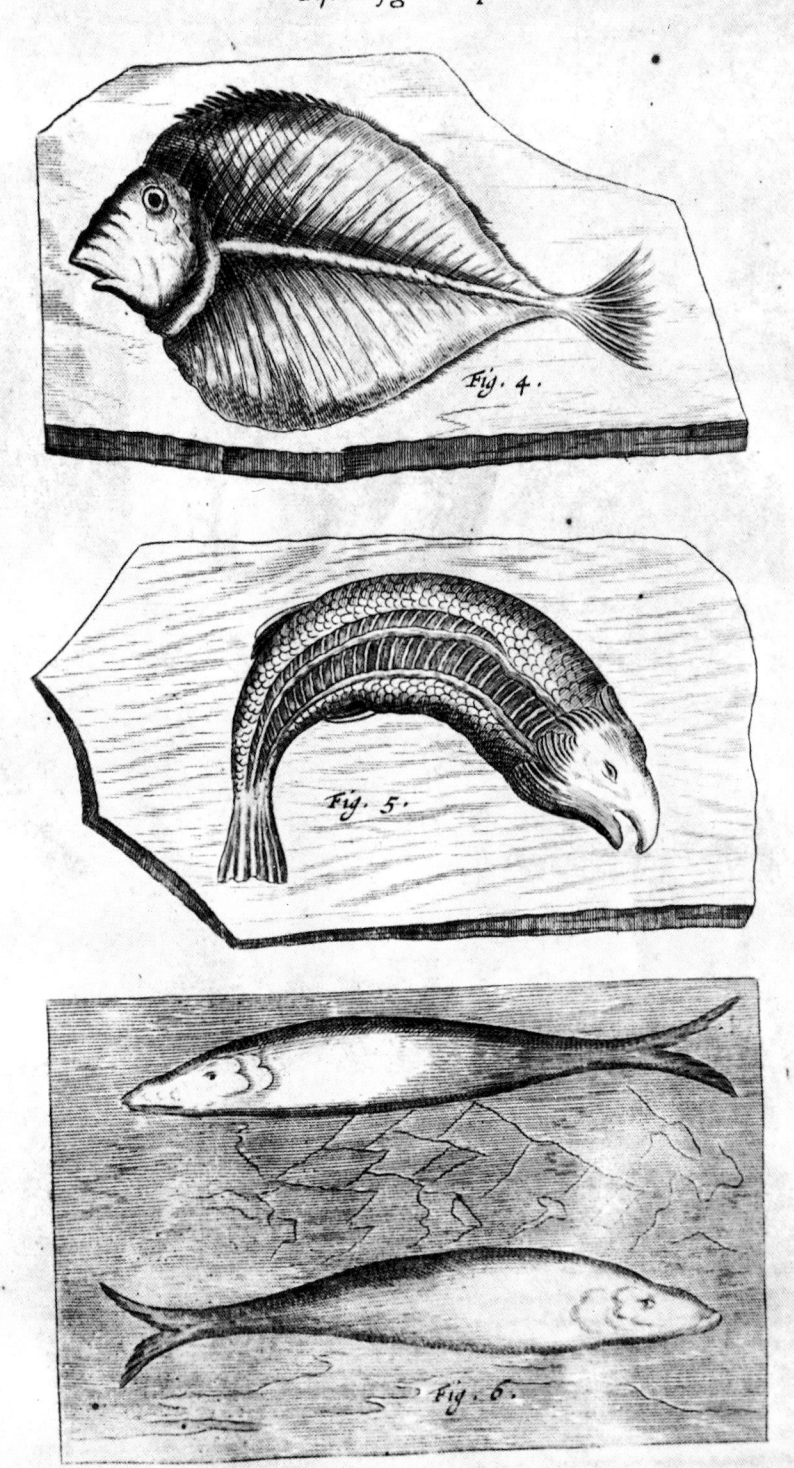

4. *Ex* Rhomborum *genere piscem refert.*
5. Auratæ *piscis formam exprimit; pariter sicuti cæteri intra saxum in lapidem conversi.*
6. *Duos pisces veluti aquæ innatantes exhibet, qui in Museo* Aldrovandino *spectantur.*

TAB. IV.

next fifteen years. This lucrative association marked his entree into an international learned community of far more expansive range than the clerical circles and carefully watched gentlemen's academies of Italy. It also called attention to the progressive decline of the Italian printing industry under the continuing scrutiny of the Holy Office, and to Pope Alexander VII's awareness of that decline.

In the first of the two immense folios that make up the *Subterranean World*, Kircher began by reverting to his original vocation as a mathematician, using geometry to demonstrate that the earth was the center of the universe, with reasoning no more impressive than the *Ecstatic Heavenly Journey*'s halfhearted gestures toward refuting Copernicus. Quickly, however, he moved to the work's real point, which was to demonstrate the extent to which the sun, moon, and earth were irregular bodies roiled from core to surface by the constant recombination of the four elements. As evidence of this universal flux, the *Subterranean World* tracked the eruption of sunspots, the placement of lunar craters—one of which now bears Kircher's name—and formulated an early version of plate tectonics. For a man who believed that the earth was only 6,000 years old, the speed of the geological changes that raised mountains and sank continents had to be dizzyingly swift:

> That endless duration and perpetuity of the natural operations and Elements . . . cannot be imitated by any human art or industry, for they belong only to the Omnipotence of the divine art, whose parts, even though they may seem to perish, yet nothing can perish or be lost entirely according to its whole being, so long as they remain subject to the dictates of Nature's law. Hence that perpetual, tireless whirling of the Stars comes into being, ever proceeding by some constant and inviolable law over so many myriads of years, hence the motions of the Elements is ever invariable amid Nature's great variety. Hence the cycling of the Waters in and around the Earth is perpetual, and as the Wise Man testifies: "*All the rivers run into the sea; yet the sea is not full; unto the place from whence the rivers came, thither they return again* [Eccl. 1:7]; the Sun evaporates the Sea by its vapors, the vapors are dissolved in rain, whence they are extracted, and are soon restored."

The fiery core of Kircher's earth, the furnace, or *pyrophylacium*, that heated the magma of the world's volcanoes, had no room for a traditional Hell, but only for raging chemical transformations. His was a surprisingly modern world, one that implied a surprisingly modern theology. Sin and its punishment were no longer defined in strict physical terms, but as metaphysical estrangement from God. Creation, moreover, was no longer static but rather subject to a fixed law of incessant change.

Together with this incessant change, however, Kircher also perceived a world of infinite fertility, in which the very stones were moved by God's love:

> Just as the human heart is visibly captured by the urgings of an inundation of love without visible chains, so too [the magnet], captured by its love for its desired iron, is penetrated by the ties of an invisible embrace, so that they cannot be separated from one another except by force.

*This lucrative association marked his entree into an international learned community of far more expansive range than the clerical circles and carefully watched gentlemen's academies of Italy.*

Chemical compounds also gave rise to their own variety of being, which he located somewhere between the degree of organization that defined life and the randomness of accident. From watching the growth of crystals, Kircher believed that salt, sulphur, and mercury brought about their own version of generation, and although he distinguished this form of mineral life from the vital spark that made plants and animals move and grow, he nonetheless regarded it as a distinct mode of being. His world was a place of boundless creative force. He believed that volcanoes replenished themselves when rain carried off the salts in solidified lava, leaching them back into the deep recesses of the earth. Insects, to his mind, arose from mammal dung, and orchids sprang from semen dropped on the ground. He applied a Greek term, *panspermia*, to this general principle of fertility; the word meant "universal seed," or as he himself defined it, "a seminal or spermatic power, by whose qualities and efficacy things appear by natural propagation from what has perished." Potentially, therefore, each element of nature had the power to propagate itself, and by this means the world, however corruptible from moment to moment, preserved its immortality.

The sovereign principle that governed Kircher's doctrine of *panspermia* was change, to which he frequently applied the term "vicissitude." It was another echo from the writings of Giordano Bruno, although only Kircher applied such a doctrine so trenchantly to the workings of the earth. For Bruno the consistent uncertainty of the phenomenal universe was counterbalanced by a universal, all-penetrating world soul. Kircher's *panspermia* served much the same purpose, but works like the *Subterranean World* are so filled with eruptions of praise from the Psalms that no censor could ever have doubted the good Jesuit's Christian sincerity.

Like Bruno before him, however, Kircher had no intention of standing still in rapturous praise of Creation. Each in his turn, both renegade philosopher and Jesuit divine proposed an active mental discipline by which human beings could—indeed should—face the sheer immensity of nature, and for both the first step in that discipline entailed reviving the ancient Greco-Roman art of memory. Bruno's purpose in cultivating memory had become, by the end of his career, abstractly philosophical. Kircher, on the other hand, whose management of memory was conditioned by the *Spiritual Exercises* of Ignatius, never lost sight of a fundamental religious purpose behind his mnemonic exertions. His international reputation only served to remind him that he should strive perpetually to act as the embodied microcosm of his Church and his order. As said in the *Subterranean World*:

> But the Conversions of the Terrestrial Globe are so large and so horrible that they lay bare both the infinite power of GOD and the uncertainty of human fortune, and warn the human inhabitants of this Geocosmos that as they recognize that nothing is perpetual and stable, but that all things are fallible, subject to the varying fates of fortune and mortality, they should raise their thoughts, studies, soul and mind, which can be satisfied by no tangible object, toward the sublime and eternal Good, and long for GOD alone, in whose hands are all the laws of Kingdoms, and the boundaries of universal Nature.

Once again, however, Kircher's real speculations about the use of mind and memory traveled deep, to emerge in another large folio book that represents, in effect, the philosophical sequel to the *Subterranean World*. This latest step in the Jesuit's intellectual itinerary was the *Great Art of Knowing*, published in Amsterdam by Jansson and Weyerstraet in 1669. Through this "Great Art," Kircher contended, all the fundamental tenets of Christianity could be reduced to a logical system expressed in twenty-seven symbols and communicated with the clarity and precision of syllogism. His long search for a universal language had finally induced him to devise one of his own, made up of letters of the alphabet and a series of symbols he devised himself; they were, to all intents and purposes, a set of modern hieroglyphs. These twenty-seven symbols could be manipulated in two ways to yield their ranges of meaning. The first manipulative procedure, the Combinatory Art, elaborated on a kind of memory training that was first devised by the medieval Catalan mystic Ramon Llull in the early fourteenth century. The practitioner of Llull's technique memorized information by imagining ideas and facts as neatly stored on a set of concentric wheels. As Kircher pointed out, however, subsequent scholars "had edited vast commentaries on [Llull]," to make his work more accessible. These commentators, as Kircher stated, included one "Jordanus," whose last name, carefully omitted, was Bruno.

A second, more innovative aspect of Kircher's "Great Art of Knowing" involved analogy, putting the world into order by drawing connections among its various parts. Kircher chose his examples carefully, to show the art's applicability to missionary work:

> As a man, Christ had being like a stone, life like a plant, senses like an animal, reason like a man. He understood like an angel, and worked all things in all things and through all things, like God.

But the frontispiece to the *Great Art of Knowing* shows that Kircher's art aimed at something far more elusive than preaching the Gospel to the heathen. In this elaborate engraving, Divine Wisdom sits enthroned before the sun, flanked by the senses of sight and hearing, and floating above a seashore; rivers flow into the sea while a volcano spews its magma into the atmosphere. The Great Art of Knowing, in other words, forges a bond between the phenomenal world and the celestial one, reconciling the vicissitudes of nature with eternal truths. Its practitioner begins to understand what it is to exist in the universe.

Ramon Llull said in 1308 that his Combinatory Art allowed its adepts to know and love God. Giordano Bruno's version of the Llullian art promised that through trained memory, humanity would discover untold new powers to interact with a universe opened out into infinity: "what then can we not understand, remember, and do?" Kircher's version of the art, more recognizably orthodox, perhaps, was scarcely less ambitious: it led to "Wisdom, the explorer of the loftiest matters, who, passing far beyond the limits of human joy, joins her own [voice] to the Angelic Choruses, and borne before the Ultimate Throne of Divinity, makes them consorts and possessors of Divine Nature."

On the title page of the *Great Art of Knowing*, and again on its first page of text,

Kircher quoted Plato's remark that "Nothing is more beautiful than to know everything." For decades Athanasius Kircher devoted himself to knowing everything, from the life of the tiniest worms to the outermost reaches of an infinite heaven. Mindful of human limitations, he succeeded as consistently as anyone in his own era at pushing back the boundaries of the senses, reversing the direction of time, and drawing connections—not only analogies between disparate phenomena, but bonds between people. He produced many other publications in addition to those mentioned so far: books on Noah's Ark (*Arca Noe*, 1675), on the Tower of Babel (*Turris Babel*, 1679), on mummies (*Sphinx mystagoga*, 1676), on the local antiquities of the Roman countryside (*Latium*, 1671; *Historia Eustachio-Mariana*, 1665), on magnetism (*Magneticum naturae regnum*, 1667), and on the effects of volcanic ash on the inhabitants of Naples when Vesuvius erupted (*De prodigiosis crucibus*, 1661).

The staggering volume of this output was made possible in part because Kircher repeated himself from book to book, often for pages on end. Specious arguments appear alongside serious investigations, reflex conformity to doctrine alongside near-heresy. Many of his more discriminating readers began to rue the effort it took to determine which was which. By the end of his career, he had both an international reputation and a diehard set of detractors. To his student Johann Stephan Kestler, he was "great Kircher, the most miraculous mystagogue of Nature, the great magician." To Melchior Inchofer, he was the itinerant Egyptian, spouting trifles about the moon.

It is scarcely easier to assess Kircher today. Because he came as close as anyone in his time to knowing everything, much of what he knew was wrong. Spontaneous generation of insects does not occur from mammal dung. All languages do not descend from Hebrew. The world's great mountain ranges do not conceal vast *hydrophylacia* that feed all the planet's streams and rivers. Still, Kircher's instincts could be extraordinarily sharp, and he often intuited what we would now claim to know with scientific precision. His infinite cosmos, with its chemically complex bodies, is a place we can recognize, along with his fiery sun, pockmarked moon, and permeable earth. We know what he suspected: that plague is borne by a bacillus too small for his *Smicroscopium* to detect. Ironically, however, this belief that the Black Death was spread by contagion also led Kircher to accept one of the most horrific episodes to mar its outbreak of 1631: in Milan (as Alessandro Manzoni would later recount with chilling vividness) people called "anointers," *untori*, were accused of daubing surfaces like walls and benches with pestilent grease. A few unfortunates were rounded up, tortured into confession, and executed; according to Kircher's theory of contagion, the tale of their crimes and punishment made perfect sense. The glimmer-

*[H]e succeeded as consistently as anyone in his own era at pushing back the boundaries of the senses, reversing the direction of time, and drawing connections—not only analogies between disparate phenomena, but bonds between people.*

Dialectica   Rhetorica   Physica   Medicina   Mathesis

ings of enlightenment did not necessarily produce the full reign of reason.

More pointedly, Kircher's need to express himself in guarded language and elaborate codes has made it difficult to know with certainty what he really meant to say. The elaborate art of dissembling by which he survived in an age of censorship took its toll on Jesuit science, on the Catholic press, and on Kircher as an individual. The author of the entry on Kircher in the 1909 American edition of the *Encyclopaedia Britannica* surveyed his decades of work and reached a sadly damning conclusion:

> Kircher was a man of wide and varied learning, but singularly devoid of judgment and critical discernment. His voluminous writings . . . often accordingly have a good deal of the historical interest which attaches to pioneering work, however imperfectly performed; otherwise they now take rank as curiosities of literature merely.

In fact, however, as this essay has endeavored to show, Athanasius Kircher may have furthered the scientific enterprise more effectively than has been recognized, by "subterranean paths" of veiled allusion that nonetheless allowed him to discuss some of the most controversial ideas to animate his troubled times. Meanwhile, the sheer beauty of his picture books and the force of his industry bore witness to the compelling magnificence of a world whose scale and complexity he acknowledged as few others in his day even dared to imagine.

1     Tab. XIII     2     52

# The Ecstatic Journey: Athanasius Kircher in Baroque Rome

To his student Johann Stephan Kestler, Athanasius Kircher (1602–1680) was "great Kircher, the most miraculous mystagogue of Nature, the great magician." To the French scholar Pierre Gassendi, another contemporary, he was "a man from the Society of Jesus of the greatest erudition . . . a great expert in the mysteries of the Hieroglyphs." To the authors of the 1909 American edition of the *Encyclopaedia Britannica*, on the other hand, "Kircher was a man of wide and varied learning, but singularly devoid of judgment and critical discernment." Called to Rome as professor of mathematics at the Jesuit College in 1635, just after Galileo's trial and condemnation by the Roman Inquisition (1633), Kircher was given two impossible missions by his superiors: to reconcile Church doctrine with experimental method, and to interpret Egyptian hieroglyphs. Over the course of the next forty-five years, in more than forty books, he addressed virtually every topic of interest in the seventeenth century, from the findings of the microscope to the structure of the cosmos, from the decipherment of ancient Egyptian to a symbolic logic of his own invention. A master communicator who used sound effects, the magic lantern, and practical jokes to spice his live presentations, and curator of a museum that was one of Rome's notable attractions, Kircher was also a consummate exploiter of the power of the printed book, as this exhibition aims to show.

1. Photograph reproduced from Georgius de Sepibus (fl. 1678). *Romani collegii Societatus [sic] Jesu musæum celeberrimum.* Amsterdam: Ex officina Janssonio-Waesbergiana, 1678.
Presented in Memory of Cora B. Perrine

Athanasius Kircher, shown here in a portrait frontispiece, was born in Geisa, Germany, in 1602. Educated in Jesuit schools, he took his vows as a member of the Society of Jesus at Paderborn in 1620. As a result of the civil strife between Protestants and Catholics known as the Thirty Years' War, Kircher wandered throughout Germany for more than a decade. During this time he was ordained as a priest (in Mainz in 1628); taught mathematics, Hebrew and Syriac; published his first book on magnetism; and began his lifelong study of Egyptian hieroglyphs. The Swedish invasion of Germany in 1630–31 induced Kircher to flee to the safety of Avignon together with his student and colleague, the Jesuit scientist Gaspar Schott. In France he met the scholar Nicolas Claude Fabri de Peiresc, who in turn recommended him to Cardinal Francesco Barberini, nephew of the reigning pope, Urban VIII. In 1634 Kircher received a summons to replace Johannes Kepler as mathematician to the Hapsburg court in Vienna, but decided to visit Rome en route. By the time he arrived in the Eternal City in 1635, his orders had changed: through the offices of Cardinal Barberini, he had been appointed instead as professor of mathematics at the Jesuits' Collegio Romano (Roman College). Aside from a trip to Southern Italy, Sicily, and Malta in 1637–38, he would remain in Rome and its environs until his death in 1680, turning his attentions to nearly every aspect of learning, publishing over forty books, and assembling an ever-growing museum on the premises of the Collegio Romano. His initial reputation for supreme erudition was somewhat compromised in later years as his theories on a whole range of subjects came under attack from religious and scientific adversaries, but Kircher also produced a loyal cadre of students, including fellow Jesuits Gaspar Schott and Filippo Buonanni, who carried on his researches into subsequent generations.

CAT. I

# Athanasius Kircher, S.J.

> Plato said, "Nothing is more divine than to know everything," sagely and elegantly, for just as Knowledge illuminates the mind, refines the intellect and pursues universal truths, so out of the love of beautiful things it quickly conceives and then gives birth to a daughter, Wisdom, the explorer of the loftiest matters, who, passing far beyond the limits of human joy, joins her own to the Angelic Choruses, and borne before the Ultimate Throne of Divinity, makes them consorts and possessors of Divine Nature."
>
> ATHANASIUS KIRCHER, *Ars magna sciendi* [1669], P. *3 RECTO

FOR DECADES ATHANASIUS KIRCHER DEVOTED himself to knowing everything, from the life of the tiniest worms to the outermost reaches of an infinite heaven. Mindful of human limitations, he succeeded as consistently as anyone in his own era in pushing back the boundaries of the senses, through new scientific instruments and new habits of thought. He worked under extraordinary restrictions, imposed on his writings and teachings by the standard curriculum of his order, the Society of Jesus, and by the unpredictable judgments of censors and inquisitors in Counter-Reformation Rome. Within these limitations, Kircher found his own way to express new ideas. He was fortunate enough to have (and shrewd enough to retain) the protection, based on scholarly respect, of influential churchmen like Cardinal Francesco Barberini and Pope Alexander VII, whose approval allowed him a certain degree of freedom to air his sometimes unorthodox ideas. He also worked, as did his contemporaries—Catholic, Protestant, and Jewish—through symbols, metaphors, allegories, and riddles, declaring that the most important truths should not be subjected to outright exposure. With all the strange variety of the physical world, the mystery essential to Catholic theology, and the layers of symbolic allusion in which he embedded some of his most significant points, not to mention the sheer scope of his interests, Kircher's works, like Kircher the man, represent a massive challenge. Yet, deliberately provocative and often beautiful, the books retain their fascination long after most of their ideas have become obsolete, timelessly appealing for their lively response to the pressing problems posed by new ideas in a traditional society.

Kircheriana Domus naturæ artisq, theatrum
Par cui vix alibi cernere poſſe datur.
*AMSTELODAMI.*
Ex officina Janſſonio-Waesbergiana Anno MDCLXXVIII.

CAT. 2

2. Photograph reproduced from Georgius de Sepibus (fl. 1678). *Romani collegii Societatus* [sic] *Jesu musæum celeberrimum*. Amsterdam: Ex officina Janssonio-Waesbergiana, 1678.
Presented in Memory of Cora B. Perrine

The vast range of Kircher's activities is conveyed in this engraved title page, showing him in his *Musaeum*—"This workshop of Art and Nature, this treasury of the Mathematical Disciplines, this Epitome of practical philosophy, the Musaeum Kircherianum. . . ."

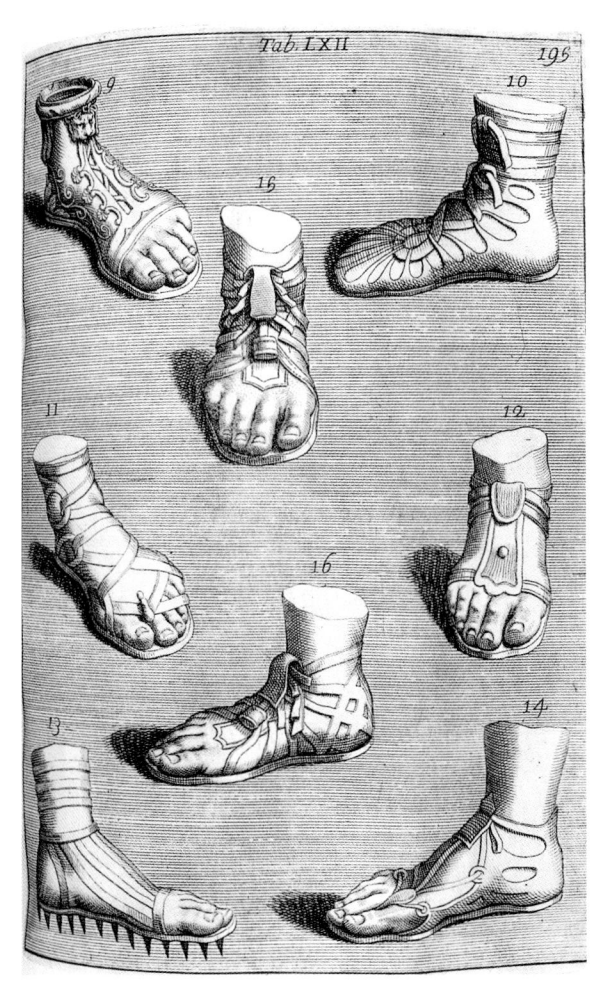

CAT. 4

3. Georgius de Sepibus (fl. 1678). *Romani collegii Societatus* [sic] *Jesu musæum celeberrimum*. Amsterdam: Ex officina Janssonio-Waesbergiana, 1678.
Presented in Memory of Cora B. Perrine

From 1665 on, the Amsterdam firm of Jansson and Weyerstraet published a lavish series of Kircher's books, both in their original Latin and in French, German, and Dutch translations. This introduction to Kircher and his thought uses the *Musaeum* as its guiding theme. It is not, however, a catalogue of the collection itself, but rather an introductory anthology of Kircher's works, to encourage purchase of the complete volumes available from the Jansson and Weyerstraet press.

4. Filippo Buonanni (1638–1725). *Musæum Kircherianum*. Rome: Typis Georgii Plachi, 1709.

The distinguished Jesuit scientist Filippo Buonanni assumed care of the *Musaeum* at Kircher's death, in 1680. By the time Buonanni drafted this catalogue, the collections had already begun to be dispersed. Still, despite continuing losses, the *Musaeum*'s holdings were important enough in the early twentieth century to provide seed collections for several national museums in Rome, including the Villa Giulia museum of Etruscan antiquities, the Luigi Pigorini Ethnographic Museum, and the museum of classical antiquities originally housed in the Baths of Diocletian and now relocated to the Palazzo Massimo. Many of Kircher's curiosities, including his wooden models of obelisks, are still preserved in the laboratories of the Liceo Visconti, the high school that occupies much of the site of the former Collegio Romano.

# The City of Rome

SHRUNK TO A FRACTION of its former size in the Middle Ages, Rome began a long process of rebuilding at the beginning of the fifteenth century. By Kircher's time, the city had gradually expanded within the compass of its ancient walls (built by Emperor Aurelian in ca. 270 C.E.). Nonetheless, for their imposing bulk and their hold on the imagination, the ruins of the ancient city continued to dominate the Roman landscape. Every achievement of modern Rome was inevitably measured against the massive legacy of the past.

5. *Roma antiqua*. [Rome]: Gotfridus de Scachijs, [ca. 1620–35]. Engraving added to *Speculum Romanae magnificentiae*. Rome: Antonio Lafreri, ca. 1544–77.

This engraving provides a view of ancient Rome through Baroque eyes, when modern buildings were just beginning to fill out the perimeter of the ancient walls, and ruins loomed larger in the landscape than they do today.

6. *Trophea Marii*. Rome: Claudij Duchetti, [ca. 1581–86]. Engraving added to *Speculum Romanae magnificentiae*. Rome: Antonio Lafreri, ca. 1544–77.

Originally displayed in niches on an elaborate ancient Roman fountain, these marble trophies (heaps of captured enemy armor)—fancifully ascribed to Julius Caesar's mentor, the popular general Gaius Marius—were eventually removed for display on the balustrade of the Campidoglio, the Capitoline Hill, where they still stand today.

7. Pirro Ligorio (1510–1583), artist; Ambrogio Brambilla (fl. 1579–1599), engraver. *Circi maximi acvratissima descriptio*. Rome: Clavdii Dvcheti, 1581. Engraving added to *Speculum Romanae magnificentiae*. Rome: Antonio Lafreri, ca. 1544–77.

Many of the Egyptian obelisks brought to Rome were displayed along the central spines of the circuses in which the Romans raced chariots, as in this lively reconstruction of the Circus Maximus. The emperor Domitian (reigned 82–96 C.E.) also put an obelisk in the temple of the Egyptian deity Isis, whence it was removed by the emperor Maxentius (reigned 306–312 C.E.) to his circus on the Appian Way. It was finally relocated in 1648 to the site of the old Circus of Domitian, Piazza Navona, to form the centerpiece for Gianlorenzo Bernini's Fountain of the Four Rivers. Athanasius Kircher translated the hieroglyphic inscriptions on this well-traveled obelisk into Latin as part of the fountain's design and published his findings in a large study dedicated to Pope Innocent X Pamphili, *The Pamphili Obelisk* (*Obeliscus Pamphilius*).

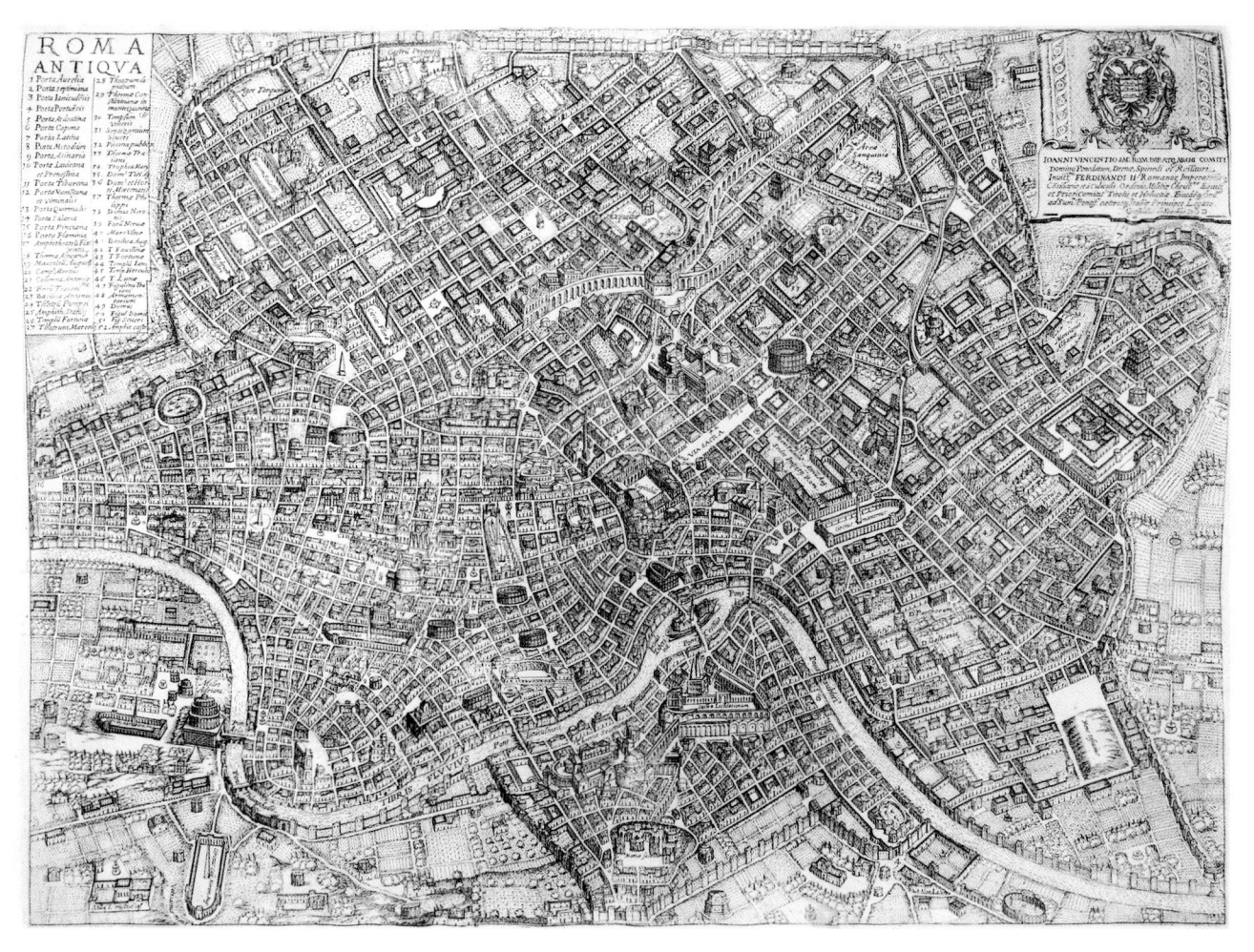

CAT. 5

# The Counter-Reformation and the Jesuits

MARTIN LUTHER WAS ONLY the most impatient of the many early sixteenth-century Christians who believed that their Church sorely needed reform. As the wealthy states of Northern Europe increasingly fell under the sway of Protestant movements, Pope Paul III issued a bull in 1542 calling for a general council of the Roman Church to meet at Trent, in the Italian Alps:

> Whereas we deemed it necessary that there should be one fold and one shepherd, for the Lord's flock in order to maintain the Christian religion in its integrity, and to confirm within us the hope of heavenly things; the unity of the Christian name was rent and well-nigh torn asunder by schisms, dissensions, heresies. . . . Then, recalling to mind that our predecessors, men endowed with admirable wisdom and sanctity, had often, in the extremest perils of the Christian commonweal, had recourse to ecumenical councils and general assemblies of bishops, as the best and most opportune remedy, we also fixed our mind on holding a general council.

By the time it closed in 1563, the Council of Trent had issued a series of decrees about the right to interpret Scripture, the meaning of the sacraments, the censorship of books, and the proper purpose of religious art. But rather than defuse the pressures of Protestantism, the council only made the schism more definite. At the same time, the internal reform movement of Roman Catholicism gave rise to new religious orders, one of the most successful of which was Kircher's own, the Society of Jesus, founded by Ignatius Loyola and formally recognized by Pope Paul III in 1540.

Ignatius hoped to create an army of priests, each of them prepared to move anywhere in the world to preach the Gospel and to embody it. The deceptively simple procedures of his *Spiritual Exercises* produced, through rigorous memory training, a remarkably flexible mental discipline. Together with the society's early emphasis on an educational mission, this shared spiritual instruction worked with surprising rapidity to forge an order of sophisticates, adaptable to the courts of Europe, Asia, and the Americas as well as to the hospices of the urban poor. To further their ability to maneuver, the Jesuits allowed themselves unusual freedom to adapt to local conditions.

8. Petrus Suavis Polanus [Paolo Sarpi (1552–1623)].
*Historiæ Concilii Tridentini.* Frankfurt: Apud
Godefridum Tampachium, 1621.
Ernst Wilhelm Hengstenberg Collection

For the Venetian Servite friar Paolo Sarpi, the reforms enacted by the Council of Trent reflected not God's will working through the Holy Spirit, but the basest of human wranglings. He therefore published his *History of the Council of Trent* (*Historiae Concilii Tridentini*) under the pseudonym of Petrus Suavis Polanus, and used a Protestant press:

> My proposal is to write the history of the Council of Trent.... I will tell the causes and conduct of a Church convention over the course of twenty-two years, to differing purposes and by differing means, who promoted and solicited it and who obstructed and dissented from it, and how it was convened and dissolved on and off again for another eighteen years: always publicized to various purposes, [a council] that achieved a form and a result entirely contrary to the plans to those who brought it about, and to the fears of those who disrupted it. Clear testimony to resign your thoughts to God, and not to trust in human prudence. For this Council, desired and brought about by pious men in order to reunite the Church, which had begun to split apart, so confirmed the schism and solidified the two sides that it made the differences irreconcilable; and manipulated by Princes to reform the order of the Church, it has caused the greatest distortion that has ever occurred in the life of Christianity.... It would not be inappropriate to call it the *Iliad* of our century, in the narration of which I shall follow straight after the truth, as I am not possessed by any passion that can make me go astray. (PP. 1–2)

9. Jean Barbault (1718–1762), artist; Freicenet, engraver.
*Veduta del Collegio Romano.* Rome: Bouchard e
Gravier, [1763]. Engraving added to *Speculum
Romanae magnificentiae.* Rome: Antonio Lafreri,
ca. 1544–77.

An eighteenth-century engraving of the Jesuits' Collegio Romano, built in 1582 on the site of the ancient Roman Temple of Isis. Designed by Francesco Valeriani, S.J., and Bartolommeo Ammannati, it was arranged around a series of courtyards, ranging from the elegantly formal to more private quarters around the kitchen, refectory, and pharmacy. Its vast library contained forbidden books as well as those that met with Church approval. Eventually the complex incorporated the immense church of Sant' Ignazio, designed by Orazio Grassi, S.J., one of Kircher's predecessors in the Collegio's chair of mathematics. From 1651 onward, the Collegio building also housed Kircher's *Musaeum*. This stately engraving shows none of the bristling array of scientific instruments that must have studded the Collegio's roof and hung from its windows, not to mention the windowsills that Kircher loaded with substances in various states of decay.

10. Saint Ignatius of Loyola (1491–1556). *Exercicios espirituales.* Seville: Imprenta de los Recientes, [17—].

This eighteenth-century Spanish edition of the basic Jesuit text, a four-week guide to meditation on Christian doctrine, is open to the First Week, Fifth Exercise:

> This is a meditation on Hell. It contains a preparatory prayer, two preludes, five points, and a colloquy.
> Preparatory Prayer: This will be as usual.
> First prelude: This is the representation of place. Here it will be to see in imagination the length, breadth, and depth of hell.
> Second prelude: I will ask for what I desire. Here it will be to ask for a deep awareness of the pain suffered by the damned, so that if I should forget the love of the Eternal Lord, at least the fear of punishment will help me to avoid falling into sin.
> First Point: To see in imagination the great fires, and the souls enveloped, as it were, in bodies of fire.
> Second Point: to hear the wailing, the screaming, cries, and blasphemies against Christ our Lord and all his saints.
> Third point: to smell the smoke, the brimstone, the corruption, and rottenness.
> Fourth point: to taste bitter things, as tears, sadness, and remorse of conscience.
> Fifth point: with the sense of touch to feel how the flames surround and burn souls.
> Colloquy: Enter into a colloquy with Christ our Lord. Recall to mind the souls in hell . . . Conclude with an "Our Father."

(Translated by Anthony Mottola, S.J., *The Spiritual Exercises of Saint Ignatius* [New York: Doubleday, 1964], P. 59)

*Veduta del Collegio Romano*

CAT. 9

11. *Acta Concilii Tridentini*. Antwerp: Excudebat Martinus Nutius, 1546.

This little book provides a record of the proceedings from the first four sessions of the Council of Trent, held in 1545–46. The list of participants displayed here includes a future pope, Marcello Cervini, among the high-minded reformers like Reginald Cardinal Pole, whose purposes, according to Paolo Sarpi, were eventually reversed by the council's increasingly dogmatic tendencies. The presence of the Swedish archbishop Olaus Magnus represents a final, and unsuccessful, attempt to keep that country Catholic.

12. Photograph reproduced from Georgius de Sepibus (fl. 1678). *Romani collegii Societatus [sic] Jesu musæum celeberrimum*. Amsterdam: Ex officina Janssonio-Waesbergiana, 1678.
Presented in Memory of Cora B. Perrine

Kircher wrote one of the earliest explanations of how to build and use a magic lantern. Here he projects the image of a soul in Purgatory, to whose flames the candlelight of the magic lantern must have lent a realistic flicker.

13. Photograph reproduced from Athanasius Kircher (1602–1680). *Toonneel van China, door veel, zo geestelijke als weereltlijke geheugteekenen....* Amsterdam: Johannes Janssonius van Waesberge, 1668.
Courtesy of Getty Research Institute for the History of Art and the Humanities

Wistfully compiled from the reports filed in Rome by Jesuit missionaries in Asia, Kircher's *China Illustrated in Monuments* honors four members of the society on this elaborate engraved title page: Ignatius Loyola, the order's founder; Francis Xavier, one of the order's seven founding members, sent by Ignatius as a missionary to India and Japan; and the two most successful of the many missionaries to China, Matteo Ricci (1552–1610), founder of the Chinese Mission, and Johann Adam Schall von Bell (1591–1669), whose forty-seven years in China culminated with his appointment as director of the National Board of Astronomers. Both Ricci and Schall von Bell adopted Chinese dress, as the order permitted them to do.

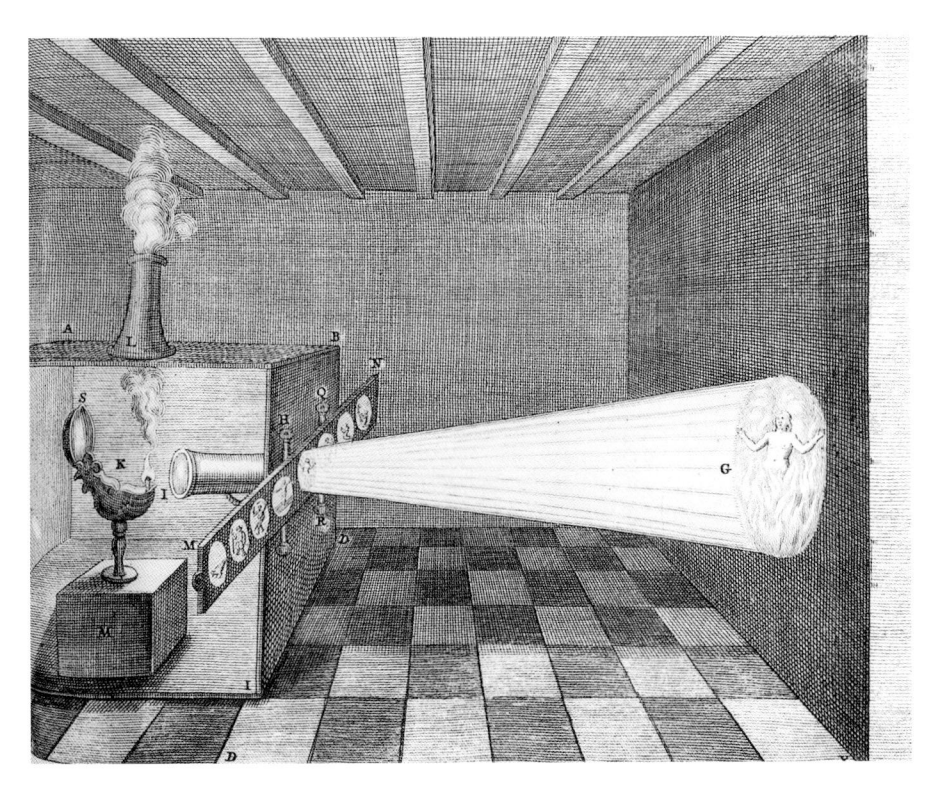

CAT. 12

## Antiquarianism and Science

FOR KIRCHER AND HIS contemporaries, the world's wonders included the remnants of past societies as well as the works of nature and the exotic variety of human culture. In Rome itself, special interest attached to the early days of Christianity; imposing basilicas and humble catacombs both bore witness to the life of the early Church. Although Early Christian art and architecture struck seventeenth-century eyes as aesthetically inferior to the works of the Roman Empire, its spiritual fervor earned the praises of natives and pilgrims alike. Sixteenth- and seventeenth-century popes redesigned the very shape of the city's streets and squares to make the experience of Rome one of constant pilgrimage: through time, through space, through stages of Christian faith. Church historians like Cesare Baronio, Antonio Bosio, and Giovanni Severano bolstered their accounts of the early days of their religion with archaeological evidence; they also thought it important to include spiritual guidelines so that archaeological investigation would increase Christian devotion. Kircher predictably included Early Christian monuments among his many interests; at the same time, he and his *Musaeum* figured among Rome's chief attractions.

14. Cesare Cardinal Baronio (1538–1607). *Il compendio de gli annali ecclesiastici*. Rome: Heredi di Giouanni Gigliotto, 1590.

One of the unfortunate offshoots of Cesare Baronio's devout Christianity was the anti-Semitism to which he gives vent on this page of his *Church Annals* (*Annali ecclesiastici*). Yet in the earliest years of the Church, the differences between Christianity and Judaism were unclear not only to the Romans who believed in neither creed, but also to the faithful themselves: there were Christian synagogues in Rome!

> This was the Hebrew nation: a nation of the most stubborn nature, inclined to superstition, greedy for gold, profuse in its pleasures, insolent in prosperity, impatient in adversity, forgetful of good deeds, blind to all rewards but the earthly, incapable of correction except by means of plague, disdainful of the customs of others, tenacious in its own rites not for the sake of piety but for rivalry. (P. 1)

15. Antonio Bosio (1575–1629). *Roma sotterranea*. Rome: appresso Guglielmo Facciotti, 1632.

To one reader of Bosio's guide to Early Christian Rome, the nudity of Adam and Eve on the late antique sarcophagus of Junius Bassus veered too close to paganism for devout eyes, and he supplied Eve with a brassiere.

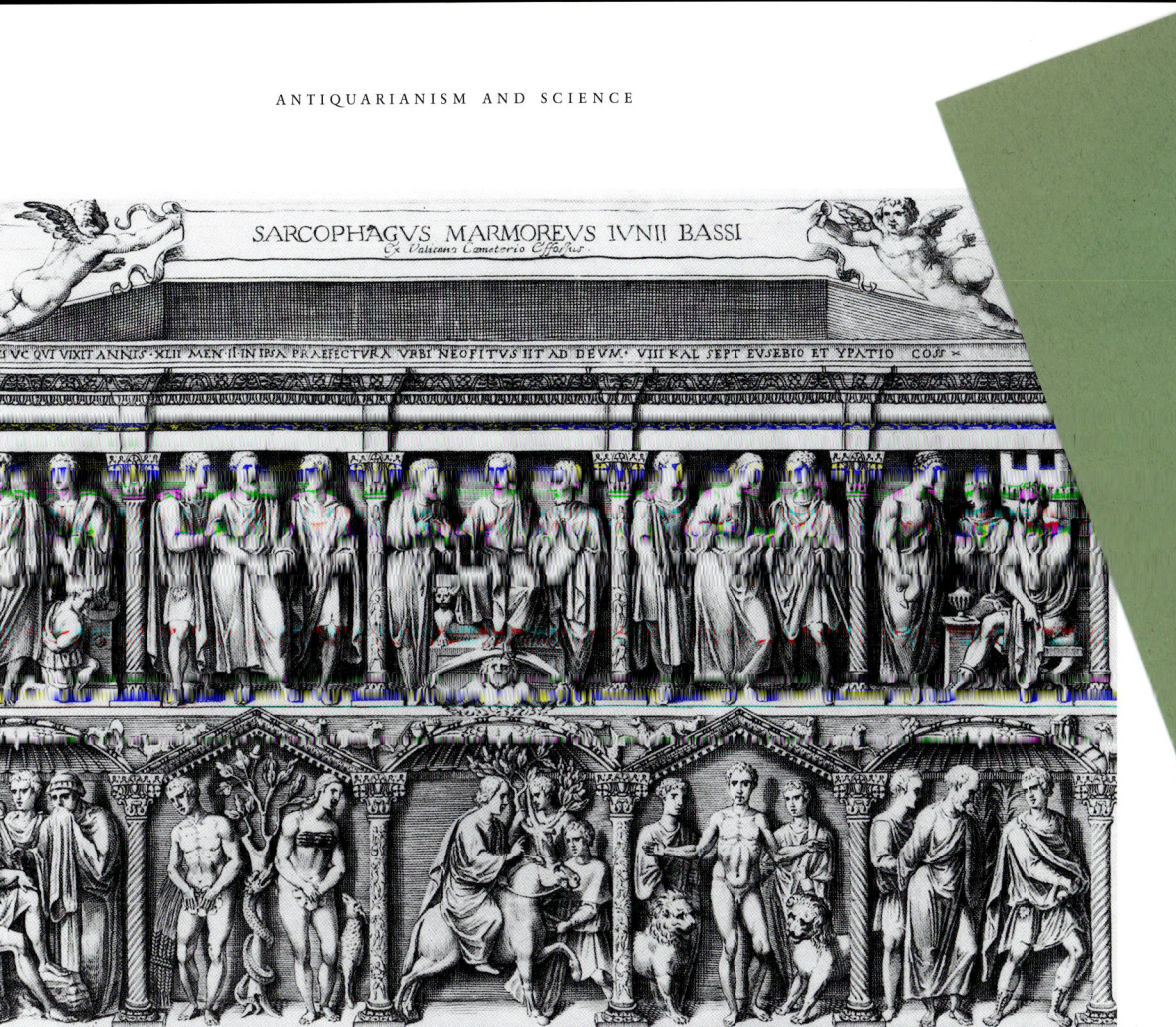

**SARCOPHAGVS MARMOREVS IVNII BASSI**
*Ex Vaticano Cœmeterio Effossus*

...ius Kircher (1602–1680). *Latium*.
...am: Apud Joannem Janssonium à
...ge & hæredes Elizei Weyerstraet, 1671.
...d from the Bequest of Louis H. Silver

...dy of Latium, the province of Rome,
...our of the area's most sumptuous modern vil-
...n to drain the malarial swamp south of Rome
...e Pomptine Marshes—a feat accomplished in
...s by Emperor Nero and again by Mussolini in

...i Severano. *Memorie sacre delle sette chiese di
...ome: Per Giacomo Mascardi, 1630.

...verano's guidebook for pilgrims bears a cen-
...nt by Antonio Bosio and a dedication to
...t mentor in Rome, Cardinal Francesco
...is instructions for visiting the obelisk
...iazza San Pietro read as follows:

When you arrive in Piazza San Pietro, where you see the Cross above the obelisk, once dedicated to the Sun in Egypt, and then dedicated by Gaius Caligula in his circus to Augustus and Tiberius, his ancestors . . . you should consider and admire how the prophecy of David has been fulfilled: *Dominus regnavit a ligno*: for divine Providence has willed it that this triumphal sign of the cross triumph over those two Emperors who had the audacity to subject the Son of God to their imperial power (that is, Augustus, who wanted to enroll Him in that universal census as his subject, and of Tiberius, who through his agent Pilate condemned Him and made Him suffer death on the Cross). Now their names and memories are trampled by that very Cross atop the obelisk that is more appropriately dedicated to the True Son of Justice, who is that same Christ, Son of God. Therefore we should venerate it with every loving devotion and adore the most precious Wood of the Holy Cross, part of which is enclosed there. (Vol. 2, pp. 10–11)

## *Scripture and Science*

GALILEO'S CONFLICTS WITH the Inquisition in 1616 and 1633 highlighted the
between observation of nature and literal interpretation of Scripture that was gr
scientific instrumentation continued to improve. By offering his own interpret
the Bible, Galileo defied a decree of the Third Session of the Council of Trent:
petulant spirits, [the Council] decrees that no one relying on his own prudenc
ters of faith and morals to the edification of what pertains to Christian doctrine
tort Holy Scripture to his own sense or against that sense which Holy Mother
holds and has held, whose place it is to judge the true sense and interpretatio
Holy Scriptures, nor shall he dare to interpret the same Holy Scripture against th
mous consensus of the Fathers, even if such interpretations were never publishe
time." Yet Galileo believed, as did Kircher, that the truth about nature would n
prove the truth of Scripture. For both of them, Psalm 19 described the sun's m
figurative language, but described the validity of God's law with literal precision

> *The heavens declare the glory of God, and the firmament sheweth His handywork.*
> *Day unto day uttereth speech, and night unto night sheweth knowledge.*
> *There is no speech nor language, where their voice is not heard.*
> *Their line is gone out through all the earth, and their words to the end of the world.*
> *In them hath he set a tabernacle for the sun,*
> *Which is as a bridegroom coming out of his chamber, and rejoiceth as a strong man to run a race.*
> *His going forth is from the end of the heaven, and his circuit unto the ends of it;*
> *And there is nothing hid from the heat thereof.*
> *The law of the LORD is perfect, converting the soul: the testimony of the LORD is sure,*
>   *making wise the simple.*
>
>                                                                              (Ps.

Philosophy is written in this great book that stands ever open before our eyes (I mean the u
verse), but it cannot be understood without first learning to understand the language and r
nize the characters in which it is written. It is written in mathematical language, and the
characters are triangles, circles, and other geometric figures; without these tools it is imposs

for a human being to understand a word; [to be] without them is to wander aimlessly through a dark labyrinth. (Galileo Galilei, *Il saggiatore* [1623], p. 25)

Whoever has the desire to pursue philosophy correctly should look to Nature's Archetype in every matter, so that by taking up Ariadne's thread in her intricate labyrinth he may keep himself safe and secure from wrong turns and deviant paths. (Athanasius Kircher, *Mundus subterraneus* [1678], 2.346)

18. Giordano Bruno (1548–1600). *De minimi existentia.* In *De monade numero et figura liber consequens quinque de minimo magno & mensura.* Frankfurt: Apud Ioan. Wechelum & Petrum Fischerum, 1591. From the Library of Joseph Halle Schaffner

[T]he Southern Italian philosopher Giordano Bruno of [N]ola was burned as a heretic in Rome in 1600, when he [re]fused to recant his belief that the universe was an infi[ni]te expanse composed of atoms, suffused by a world[-so]ul. His books were banned by the Inquisition, but they [no]netheless exerted an important influence on readers [lik]e Galileo and Kircher.

[Br]uno's dialogue on the Copernican system, *The Ash [We]dnesday Supper* (*La Cena de le Ceneri*), was one of the [fir]st works to contend that Scripture should be used for [mo]ral guidance but not as a literal account of natural [ph]enomena. Both Galileo and Kircher followed his lead [on] this issue.

*Teofilo*: Believe me, if the gods had deigned to teach us the theory of Nature as they have rendered us the favor of setting out for us the practice of morality, I would rather side with the faith of their revelations than move myself according to the certainties of my reasoning and my own sentiments. But as everyone can see clearly, in the divine books demonstrations and speculations about natural phenomena are not treated in the service of our intellect, as if they were philosophy. Instead, as a favor for our minds and affections, practice with regard to moral action is set in order through the Law. Thus, keeping this goal before his eyes, the Divine Legislator makes no effort to speak according to a truth that would do nothing to induce common people to shun evil and choose good—about this he leaves the thinking to contemplative men—and speaks to the common people in such a way that they will come to understand what is essential according to their manner of learning and speaking.
(Giordano Bruno, *The Ash Wednesday Supper* [1584], Dialogue 4)

In the philosophical poem shown here, Bruno used the same verse form as the ancient Roman philosopher Lucretius to discuss the same topic as Lucretius' *On the Nature of Things* (*De Rerum Natura*): the atomic theory of matter. Bruno's Latin dactylic hexameters are intended to capture something of the ancient writer's style as well as his basic form and content, although they do not succeed completely. Instead, Bruno's true voice spoke in his exuberant, pungent Italian vernacular:

That basic particle which cannot be split and is ageless
Fits itself in everywhere, just like the parts of the body:
Moved by you, serving your needs in a reliable order.
This is how life comes about, as we showed in detail in the *Physics*;
Here is the very source of the body's expansion and volume.
Just as a point will expand by creating a widening circle
So, too, the architect Spirit, working with groupings of atoms
Pours forth throughout the whole, everywhere, until the moment
When all the numbers play out, when the thread breaks for the body.
Then it recedes once again to the center and blends with the cosmos.
This we call Death, for we venture forth into light unfamiliar;
Only a few are granted to see that to live is to perish.
Dying, we all ascend into the genuine life.
(P. 12)

CAT. 20

19. Athanasius Kircher (1602–1680). *Ars magna lucis et vmbrae*. Rome: Sumptibus Hermanni Scheus; Ex typographia Ludouici Grignani, 1646.

Like his *Coptic Forerunner* (*Prodromus Coptus*) of ten years earlier, Kircher's *Great Art of Light and Shadow* (*Ars magna lucis et umbrae*) was censored by the Austro-Hungarian Jesuit Melchior Inchofer, who also acted as an examiner in Galileo's trial of 1633. In a book that introduces such topics as the magic lantern, the camera obscura, and, for the first time ever, fluorescence in animals and in insects like the firefly, light also has important theological significance, for Kircher was deeply inspired by the Christian Neoplatonism of the fifteenth-century Florentine philosopher Marsilio Ficino, and probably also by the more abstract Neoplatonism of Giordano Bruno. An unbroken mystical tradition connected Plato and his followers both to the philosophical movement known as Neoplatonism and to the strain of Christianity represented by the Gospel of John, the Pauline Letters, and the writings of Saint Augustine. Plato declared that the phenomenal world was an illusion of the senses, that true reality lay beyond their reach in a realm of Ideas. For him the best image of divinity was light. Saint Paul in turn would liken our present experience of the world to seeing an image in a smoked mirror ("through a glass darkly"); only death would allow us to see divinity face
to face.

> *For we know in part, and we prophesy in part.*
> *But when that which is perfect is come, then that*
>   *which is in part shall be done away.*
> *When I was a child, I spake as a child, I understood*
>   *as a child, I thought as a child: but when I became*
>   *a man,*
> *I put away childish things.*
> *For now we see through a glass, darkly; but then face*
>   *to face: now I know in part; but then shall I know*
>   *even as also I am known.*
>
> (1 Cor. 13:9–12)

Plato, as Marsilio [Ficino] says, certainly seems to have experienced a blinding flash of light more than once: when he calls it . . . an occult symbol of the Universe by which high is conjoined with low, low with middle, middle with low, and high by wondrous means. Plotinus calls [light] the secret vehicle of the universal spirit, the ape of Divinity, the laughter and joy of the heavens. . . . For what else is light, but the exuberance of God's great goodness and truth? . . . What else is [light] in Heaven if not the abundance of life among the Angels, and the expression of the power that the Platonic philosophers call the laughter of the Heavens? (Athanasius Kircher, *Ars magna lucis et umbrae* [1646], P. A recto)

The light—life, understanding, the primal unity—contains all species, perfections, truths, and degrees of phenomena. Whereas those things that occur in Nature are different from one another, contrary, and diverse, in [the light] they are like, harmonious, and single. Try, therefore, if you are able with your powers, to identify, harmonize, and unite the phenomena you perceive, and you shall not exhaust your faculties, you shall not upset your mind, and you shall not confound your memory. (Giordano Bruno, *De umbris idearum* [*On the Shadows of Ideas*] [1582], Conceptus XIII N)

20. Christoph Scheiner (1575–1650). *Rosa Vrsina, sive, Sol*. Bracciano: Apud Andream Phæum typographum ducalem, 1626–30.

The Jesuit astronomer Christoph Scheiner was Athanasius Kircher's immediate predecessor as professor of mathematics at the order's Collegio Romano, and went to Vienna in Kircher's place in 1635. His masterwork, *The Orsini Rose, or the Sun* (*Rosa Ursina, sive Sol*), privately printed by the flamboyant Roman duke Paolo Giordano Orsini, takes its title from the ultimate celestial vision of Dante's *Divine Comedy*, which is an immense ethereal rose. The work's real emphasis, however, is on the quality of Scheiner's observations with the telescope and a special device that he called the helioscope, which allowed him to observe the sun without harm to his eyes. In the engraving shown here, Jesuit astronomers are hard at work with their telescopes and helioscopes.

Scheiner's chief claim to astronomical fame rested on his work with sunspots, whose discovery he contested bitterly with Galileo: Scheiner observed them first, but thought that they were clouds in the earth's atmosphere. Eventually Scheiner realized what Galileo had believed from the outset: that the spots originated on the sun itself. Remarkably, they made these observations with relatively primitive instruments during a period when sunspot activity was unusually low, what is now known as the Maunder Minimum. For Kircher sunspots offered proof that all the heavenly bodies were subject to the same physical and chemical processes as the earth:

> In this same Body huge spots, like shadows, or counter-lights, can be observed with great wonderment: some of them, decreasing in size, reduce into the most subtle of shadows, so that, like the rest of the surface of the Sun, you can scarcely discern them without the help of an Instrument, and like a shadowy trace [*umbratile vestigium*] they will last for one or more days, whereupon they will suddenly vanish entirely, and every so often a burst of flame will occur at the same time or subsequently. (Athanasius Kircher, *Mundus subterraneus* [1678], 1.58)

21. Stanislaw Lubieniecki (1623–1675). *Theatrum cometicum*. Amsterdam: typis Danielis Baccamude; Apud Franciscum Cuperum, 1668.
John Crerar Collection of Rare Books in the History of Science and Medicine

In 1665 a comet appeared over Europe, attracting the attention of a scholarly community that revealed its truly international composition and its close interconnection when Stanislaw Lubieniecki, the exiled Polish aristocrat who compiled this *Theater of Comets* (*Theatrum cometicum*), solicited eyewitness reports of the phenomenon. Writing from Sweden, the Padua-educated physician Olof Rudbeck revealed his staunch Copernican sympathies and a pungent Latin style; Athanasius Kircher filed his report from Rome together with drawings. Lubieniecki's correspondents debated whether comets originate in sublunary space, in the region of the planets, or as far away as the fixed stars, offering their theories together with observations and drawings intended to prove the quality of their telescopes to the entire "republic of letters." The popularity of the book was immense: first printed in Amsterdam in 1667, it had undergone this second printing within a year. The Library also holds another Louvain edition, issued in 1681.

22. Galileo Galilei (1564–1642). *Il saggiatore*. Rome: Appresso Giacomo Mascardi, 1623.

The title page to *The Assayer* (*Il saggiatore*), Galileo's treatise of 1623 refuting the work on comets written in 1619 by his Jesuit rival Orazio Grassi, proudly bears two coats of arms: that of Cardinal Francesco Barberini, to whom the book is dedicated, and that of the Roman scientific academy known as the Accademia dei Lincei, "the Lynxes," to which Galileo had just been admitted as a member. Kircher voiced his own opinion of the Lincei in the *Subterranean World* (*Mundus subterraneus*): "It can hardly be denied that the wisdom of the Lyncean Philosophers of this age in scrutinizing secret motions has reached the point where they can be seen to have left all their predecessors many parasangs behind" (1.22).

CAT. 22

23. Athanasius Kircher (1602–1680). *Magnes siue de arte magnetica*. Rome: Ex typographia Ludouici Grignani, 1641.
John Crerar Collection of Rare Books in the History of Science and Medicine

Some of the images illustrating this first edition of Kircher's *Magnet, or the Magnetic Art* (*Magnes, sive de arte magnetica*) were borrowed directly from the elegant engraved plates of Christoph Scheiner's *Orsini Rose*. This illustration of a divining machine, however, was engraved to order, and it is entirely Kircher's own invention: an early version of the Ouija board, made up of wax figures suspended in water ranged around a central magnet contained within an obelisk. As a hidden crank rotated the central obelisk, the figures moved within their glass globes to spell out messages inscribed on the vessels' surfaces. Kircher loved to create devices like this in hopes that the wonders of nature and technology would vanquish belief in superstition; reality, to him, was the greatest miracle of all.

## *Urban VIII: Baroque Rome*

The pope who ruled Rome when Kircher arrived was a sophisticated Florentine, Urban VIII, the former Maffeo Barberini. Like most seventeenth-century popes, he came from an obscure family and rose through the Curia because of his talents, but once in office, he used the wealth and power of the Church to advance the status of his relatives. An urbane, genial man with a superb eye for talent, he sponsored scientists like Galileo and Tommaso Campanella; artists such as Caravaggio, Bernini, Borromini, and Pietro da Cortona; and musicians including Giulio Caccini and the castrato singer Pasqualini; as well as encouraging a variety of learned institutions in Rome, from the Dominican mother house at Santa Maria Sopra Minerva to the Sacred Congregation for the Propagation of the Faith. Under Urban VIII's leadership, the building industry continued as one of the city's chief economic resources and forever changed the face of modern Rome. The pope's chief weakness was his vanity: he was acutely conscious of his blandly handsome face, elegant manners, intellectual interests, and poetic talent. When he took offense, he retaliated with blind fury, as when he guided the Inquisition in its prosecution of Galileo, or when he mounted the expensive, futile military attack on Roman feudal barons known as the War of Castro. This was the pope into whose city Kircher arrived in 1635.

CAT. 24

observation in the mouth of the dialogue's dogmatic Aristotelian, Simplicio, and then virtually confirmed the remark's origin by having Galileo's alter ego, Salviati, exclaim, "a truly angelic doctrine!" As a strategy, it could not have been more unfortunate.

> *Simplicio*: I know that both of you, if you were asked whether God in His infinite power and wisdom could have conferred the reciprocal motion that we observe in water in some other way than by moving its container; I know, I say, that you would reply that He would be able to do so in many ways, and ways unimaginable by our intellects. I thus come to the immediate conclusion that if this be so, than it would be the supreme impudence for anyone to wish to limit and conform Divine power and wisdom to some personal fancy.
>
> *Salviati*: A wondrous and truly angelic doctrine!
> (P. 458)

25. Galileo Galilei (1564–1642). *Lettera del Signor Galileo Galilei . . . scritta alla Gran Duchessa di Toscana.* Florence [i.e. Naples: Lorenzo Ciccarelli], 1710. Bound with his *Dialogo di Galileo Galilei.* Florence [i.e. Naples: Lorenzo Ciccarelli], 1710.
Gift of Dorothy and Graham Aldis

Galileo's *Dialogue on the Two Chief World Systems* was the only one of his writings ever placed on the *Index of Prohibited Books* (*Index liborum prohibitorum*). This second edition bears a false Florentine imprint; it was actually published in Naples by the lawyer Lorenzo Ciccarelli and includes the text of Galileo's abjuration, delivered on his knees before the Inquisition on June 22, 1633. For a man who was both arthritic and inordinately proud, the painful humiliation of that moment could not have been easy to forget:

> I, Galileo son of the late Vincenzo Galilei of Florence . . . on my knees . . . with the sacrosanct Gospels before me, which I touch with my own hands, swear that I have always believed, now believe, and with God's help shall believe in future everything that the Holy Catholic and Apostolic Church holds, preaches, and teaches. But because this Holy Office has . . . intimated to me by law that I must by all means abandon the false opinion that the Sun is the center of the world and immovable, and that the earth is not the center of the world and that it moves, and that I should not hold, defend, nor teach aforesaid false doctrine in any form, whether in voice or writing . . . with sincere heart and unfeigned faith I do abjure, curse, and detest said errors and heresies. (PP. 80–[83])

24. Galileo Galilei (1564–1642). *Dialogo di Galileo Galilei . . . sopra i due massimi sistemi del mondo.* Florence: Per Gio. Batista Landini, 1632.

Galileo's small but cataclysmic *Dialogue on the Two Chief World Systems, Ptolemaic and Copernican* (*Dialogo . . . sopra i due massimi sistemi del mondo*), published with the blessing of his patron, the grand duke of Tuscany, boasted this extraordinarily fine engraved frontispiece by Stefano Della Bella. The image portrays Aristotle, Ptolemy, and Copernicus, dressed in rich robes and engaged in discussing the merits of their respective cosmological systems. The *Dialogue*'s final page contains the passage that so offended Pope Urban VIII and precipitated Galileo's conflict with the Inquisition: apparently the pope himself had suggested in conversation that there was no point in limiting the power of God by conforming His actions to any mere mortal's personal theory. Galileo put this

After having sworn off the motion of the earth according to the inquisitorial script, Galileo is rumored to have said, "It does so move" ("Eppur si muove"). The legend is untrue, but it is certainly true that three of the ten presiding cardinals refused to sign his sentence, including the pope's nephew, Francesco Barberini.

26. Photograph reproduced from Catherine Puglisi (b. 1953). *Caravaggio*. London: Phaidon Press, 1998.

When Michelangelo Merisi da Caravaggio painted this portrait at the very beginning of the seventeenth century, Maffeo Barberini was a young churchman whose sharp intellect, personal charm, and wide range of abilities held out the promise of an illustrious career. By hiring a highly successful but innovative and controversial painter as his portraitist, Barberini announced himself as a man of strong, progressive tastes. The brooding, melancholy Caravaggio, however, found little to fasten upon in his sitter's sunny personality except external characteristics: the bright eyes, ready smile, and physical energy of a vigorous, competent, but somewhat shallow man.

CAT. 26

27. Pope Urban VIII (1568–1644). *Maphaei s.r.e. Card. Barberini nunc Vrbani Papae VIII. poemata*. Paris: Typographia regia, 1642.

Before he became pope, Maffeo Barberini wrote poetry in Latin and in Tuscan vernacular, including a poem in praise of Galileo and one in praise of Mary Queen of Scots. This edition of his work is one of a long series of books issued under papal auspices from the Royal Press in Paris, testimony to Urban's close relationship with France. The poem on vanity displayed here may reflect its intelligent author's awareness of his own chief vice:

> "How useless it is to think of acquiring fame by means of poetry"
> (Quanto sia vano il pensiero d'acquistar fama col mezo della poesia)
>
> Maffeo, what's your plan? Why test your skill
> By trying to outdo the ancient sages
> Composing tuneful hymns? Where is the thrill
> In losing sleep to fill up useless pages?
>
> Fame sounds her fanfare, then she moves aside
> Like some frail bubble in a shifty stream.
> Her gifts cannot be shared; they're empty pride,
> Not nearly as eternal as they seem.
>
> Do you still love the game, the gentle error
> Of thinking you'll preserve the past in art
> And with your tuneful lyre avert Death's terror?
> Then foolish hope has carried off your heart.
>
> Good works, and not the pomp of polished speech
> Will put the gates of Heaven within your reach.
> (P. 274)

28. *Roma* [copy, ca. 1625, after Antonio Tempesta, 1593]. Engraving added to *Speculum Romanae magnificentiae*. Rome: Antonio Lafreri, ca. 1544–77.

An outstanding patron of art, Urban VIII began to reshape Rome's pilgrimage routes to celebrate Christian history and his own place in it.

# Cardinal Francesco Barberini, Kircher's First Sponsor in Rome

DESPITE HIS ORTHODOX Christian mandate, Athanasius Kircher included a wide range of thinkers among his friends. So did his first Roman patron, Cardinal Francesco Barberini (1597–1679), the nephew to whom Pope Urban VIII entrusted care of Rome's intellectual life. Barberini's eclectic collection of friends included the group of French-connected free thinkers known as *libertins* and members of the secular gentlemen's academies, like the Desiosi, "the Desirous," and the Lincei, "the Lynxes," who were beginning to pursue scientific and literary studies independent of the Church. At the same time, as cardinal, and especially as Urban's nephew, Barberini was honor-bound to uphold the authority of the papacy. He negotiated these conflicting demands with great diplomacy and high principle. He refused to sign Galileo's inquisitorial sentence in 1633, despite the fact that it meant open defiance of his irate uncle. His own cosmological beliefs almost certainly ran to the Copernican side, as this letter indicates, but it also shows that Barberini was willing to keep silent about his personal views if necessary to maintain peace:

> In your library I saw certain [globes with] Copernican Systems where the earth is mobile, made in Holland; afterwards they told me that Your Eminence ordered them taken away, so that you would not appear to follow an erroneous and condemned opinion. But it seems to me that you could preserve them for the exquisite craftsmanship with which they were made and attach a label that alerted the spectators and said "The false and lying Copernican system with a mobile earth." (Vincenzo Noghera, S.J., to Francesco Barberini, October 20, 1657; Vatican Library, MS Barb. Lat. 6488, 23 verso)

In the early 1630s, through his friends Nicolas Fabri de Peiresc and Pierre Gassendi, Cardinal Barberini came to know about a German Jesuit named Athanasius Kircher, whose work on Egyptian hieroglyphs seemed unusually promising. Soon, as Gassendi tells us, Kircher was on his way to Rome.

29. Pierre Gassendi (1592–1655). *The Mirrour of true nobility & gentility.* Translated by William Rand. London: J. Streater for Humphrey Moseley, 1657.

This biography of the free-thinking French scholar Nicolas Fabri de Peiresc was written by another scholar of similarly wide-ranging and secular interests, the Franciscan cleric Pierre Gassendi, in 1641. Gassendi's account of how the two Frenchmen met the devout German Jesuit Athanasius Kircher and recommended him to the Italian cardinal Francesco Barberini is a telling episode from the international scholarly world, termed the Republic of Letters by its participants:

> I was there when [Peiresc] wanted to summon the German Athanasius Kircher, a man from the Society of Jesus of the greatest erudition, who was then staying in Avignon. He was presented as being a great expert in the mysteries of the Hieroglyphs, for which reason, in order to be of help to [Peiresc], he sent various books as well as an image of the Mensa Isiaca. . . . And because he had in his hands the famous manuscript [of] an Arab author, Rabbi Barachas Abenephius, by whom it is said that the principles of interpreting Hieroglyphs have been handed down, therefore [Peiresc] beseeched [Kircher] to come to him, bringing the manuscript itself, a sample of his interpretation, and his notes. When [Kircher] did so, it cannot be described how ardently [Peiresc] encouraged him to finish the work that he had undertaken and prepare it for publication. The man was then called to Rome, to take over the post of the famous mathematician Christoph Scheiner, whom the Emperor wanted present in his court; Peiresc also agitated with Muzio Vitelleschi, General of the Society of Jesus, but also with Cardinal Barberini, that they add their encouragement and leave [Kircher] enough leisure to produce that work in short order. (P. 183)

30. Pierre Gassendi (1592–1655). *Institutio Astronomica: iuxta hypotheseis tam veterum, quam Copernici, et Tychonis.* Paris: Apud Ludouicum de Heuqueuille, 1647.
From the Library of Joseph Halle Schaffner

Although he is best known for his work on atomic theory, the French Franciscan Pierre Gassendi was a first-rate astronomer who was appointed royal professor of mathematics and astronomy at the University of Paris in 1645. Poor health forced him to retire shortly afterward to his native Provence, where he completed two of the books on display here: *On the Philosophy of Epicurus*, his presentation of atomic theory; and this treatise on the three competing world systems: the Ptolemaic, the Copernican, and the Tychonic. Like most seventeenth-century European scholars, he found the Copernican system the most attractive, philosophically as well as mathematically, although as an astronomer he appreciated the quality of Tycho's observations and Ptolemy's accomplishments in the age of naked-eye science. Gassendi's own observations, made from 1618 onward, form part of this volume; his greatest achievement to modern eyes was noting a transit of Mercury in 1631.

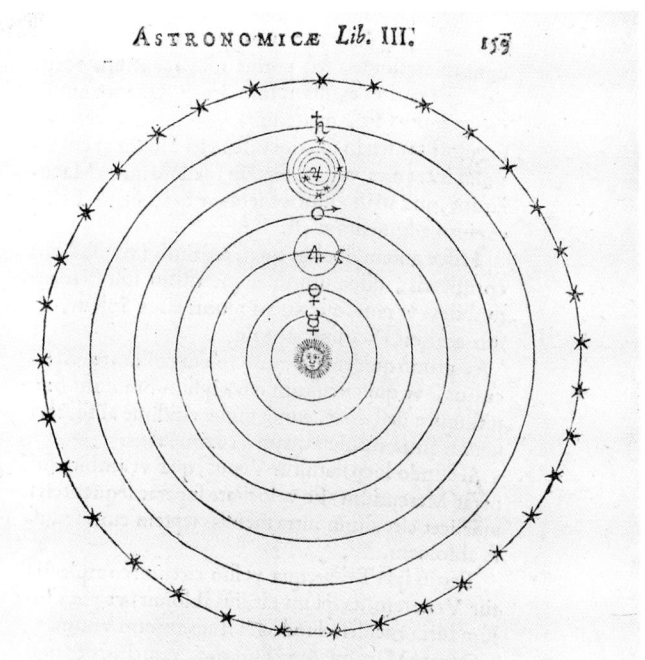

CAT. 30

31. Pierre Gassendi (1592–1655). *Animaduersiones . . . de vita, moribus, placitisque Epicuri*. Lyon: Apud Guillelmum Barbier, 1649.
   Ernst Wilhelm Hengstenberg Collection

Royal astronomer Pierre Gassendi's treatise *On the Philosophy of Epicurus* (*De vita, moribus, placitisque Epicuri*) was first published, like his *Astronomy*, in 1647, immediately after his retirement from Paris to Provence. It constituted one of the earliest and strongest statements in favor of an atomic theory of matter to emerge from the early modern period. Giordano Bruno's late sixteenth-century works on atomism, like Gassendi's, were inspired by the example of the Greek philosopher Epicurus, whose most illustrious follower, the Roman poet Lucretius, expressed his own atomistic creed in a long poem, *On the Nature of Things*. Bruno, like Lucretius, expounded his theories of atomism in Latin poetry. Gassendi, however, preferred the clarity of prose.

32. Accademia degli Informi. *Il ritratto delle virtù*. Forli: Zampa, 1697.
   Helen and Ruth Regenstein Collection of Rare Books

This pamphlet was issued by the learned society of the city of Ravenna, the "Academy of the Unformed" (Accademia degli Informi), to commemorate the dedication of a posthumous marble bust of Cardinal Francesco Barberini, who had been Legate to the province of Romagna in his lifetime. The title, *The Portrait of the Virtues* (*Il ritratto delle virtù*), aptly characterizes a man who navigated exceptionally troubled political, personal, and theological waters with competence and diplomatic grace throughout the entire course of his long life.

# Fabio Chigi/Pope Alexander VII

ON MALTA IN 1637, ATHANASIUS KIRCHER made the acquaintance of the local Apostolic Delegate and Chief Inquisitor, a nobleman from Siena named Fabio Chigi. Although one of his ancestors had been the richest man in Italy in the early sixteenth century, Chigi's family had fallen on harder times thereafter. To a great extent, like most popes of the Baroque era, he was a self-made man, would-be architect, connoisseur of art, amateur astronomer, and classical scholar, as well as a devout Christian. The friendship that he and Kircher formed on Malta would last for the rest of their lives. Chigi's service in Germany as papal nuncio in the 1640s, and then as secretary of state to Pope Innocent X, put him into contact with a broad segment of the European scholarly community, and he may have been instrumental in connecting Kircher with publishers in Amsterdam. In 1655 Chigi was elected pope, taking the name Alexander VII. For the next twelve years, Kircher was busier than at any other time in his life. At a rhythm accelerated beyond his previous, furious productivity, he prepared small manuscript pamphlets for this insatiably curious pontiff, in addition to some of his most ambitious printed works. The *Subterranean World* (*Mundus subterraneus*) (1665) was dedicated to Pope Alexander, whose presentation copy of the work is now in the Vatican Library.

33. Athanasius Kircher (1602–1680). *De specula Melitensi*. In Gaspar Schott (1608–1666). *Technica curiosa*. Würzburg: Sumptibus Johannis Andreæ Endteri, & Wolfgangi Junioris hæredum, 1664.

During his stay on Malta in 1637–38, Athanasius Kircher made a machine for the Knights Hospitallers called the "Maltese Observatory" (*Specula Melitensis*). Although the machine itself seems to have perished, Kircher's published description claims that it traced the progress of the sun, moon, and planets; tallied the dates of movable feasts on the Christian calendar; told time throughout the world; and determined auspicious times to go fishing, take medicine, or give birth. Inscribed in twelve languages (Hebrew, Chaldaean, Syrian, Arabic, Ethiopian, Coptic, Greek, Latin, Italian, French, Spanish, and German), it also combined three ideal geometric forms: circle, cube, and pyramid. In Malta, Kircher published the *Maltese Observatory* under the name of Salvatore Imbroll, Prior General of the Knights of Saint John. The original edition of the book, a small octavo printed in Naples in 1638, was already extremely rare in the seventeenth century. The entire work was included in the widely circulated *Technical Curiosities* (*Technica curiosa*), an illustrated account of some of the seventeenth century's more marvelous experiments and machines compiled by Gaspar Schott, Kircher's close friend, onetime student, and fellow Jesuit. Schott, ever solicitous of his teacher, wanted to give credit where credit was due. As he pointed out, Kircher's original edition did not require illustrations, because readers had the actual contraption before their eyes.

34. Portrait engraving of Pope Alexander VII.
    Eighteenth century.
    On loan from Professor Ingrid Rowland

A small, meticulous man with piercing eyes, few teeth, and the habit of blushing on only one side of his face, Alexander VII first struck visitors with his evident intelligence, an impression that familiarity only reinforced. A man of the highest personal integrity, he swore to do away with nepotism as pope, but he quickly discovered that for many aspects of his office, longstanding Roman custom exerted more power than personal principle. Soon many Chigi relatives were on their way to Rome, where their descendants still figure in Church and civic life. Pope Alexander's close friends also included the versatile artist Gianlorenzo Bernini. Together with his nephew Cardinal Flavio Chigi, Alexander added greatly to the Vatican Library and to the library of the University of Rome (named Biblioteca Alessandrina in his honor), as well as assembling a huge personal library and archive, which passed to the Vatican in 1923.

35. Angelo Correr (1605–1678). *Relazione della corte Romana*. Leiden: Appresso Almarigo Lorens, 1662.

Diplomatic reports such as this *relazione* by Angelo Correr, the Venetian ambassador to Rome, provided descriptions of important personalities in the various courts of Europe and synopses of the latest political struggles. Because of his illustrious diplomatic career, Alexander VII was the object of particular fascination, but an outbreak of plague in 1655, his own failing health, and the pressures of court life made his papacy less successful than it promised to be at the time of his election in 1655.

> Pope Alexander . . . does not enjoy what one would call perfect health. . . . He is left with so few teeth that if he did not compensate the loss with false ones he would mumble; indeed, with all this supplement he does not speak entirely clearly, and as for the capacity to chew, which cannot be remedied by art, he is compelled to take liquid foods and those that have no need to be broken up. He succumbs often to stomach pains, but he helps himself with exercise. Truly, the Pope's application to business could not be more intense.

> The Pope is short, with black hair that has not yet begun to gray, white of complexion, but leaden . . . he observes exquisite cleanliness in all things, but especially in eating and dress . . . he takes as much delight in dressing up as if he were in the flower of his youth. (p. 69)

36. Christian Gottfried Franckenstein (1661–1717). *Histoire des intrigues galantes de la Reine Christine de Suede*. Amsterdam: Chez Jan Henri, 1697.
    Purchased from the Andrew E. Wigeland and G. Norman Wigeland Library Book Fund

Queen Kristina of Sweden excited international attention when, in quick succession, she refused a royal marriage, abdicated her throne, converted to Catholicism, and announced her intention to move to Rome. In 1655 she made her entrance into the city, where Pope Alexander VII eagerly awaited her arrival. He was not entirely pleased with the results: to his diary, he confided his initial disappointment about her physical appearance: "non è bella" ("she is not pretty"), and the two bickered immediately about who would be entitled to the higher table at their first state dinner. To the pope's further dismay, the queen's relationship with Cardinal Decio Azzolini, the special escort he had appointed for her journey, quickly turned into a passionate intimacy that would last for the rest of their lives. This French *History of the Gallant* [i.e. romantic] *Intrigues of Queen Kristina* (*Histoire des intrigues galantes de la Reine Christine de Suede*) draws from fact as well as gossip about the Swedish queen in Italy.

37. Athanasius Kircher (1602–1680). *Magneticum naturæ regnum*. Amsterdam: Ex officina Johannis Janssonii à Waesberge & viduæ Elizei Weyerstraet, 1667.
    John Crerar Collection of Rare Books in the History of Science and Medicine

This tiny book on magnetism, *Nature's Magnetic Kingdom* (*Magneticum naturae regnum*), is dedicated to Alejandro Fabiani, a Mexican priest of Genoese ancestry. Its small size may have been designed for easy shipment to the Americas, but it is worth noting that Kircher's large books also made their way to the New World. For example, portraits of the brilliant and talented Mexican nun, Sor Juana Ines de la Cruz, show her in her library, posed before a collection in which Kircher's works enjoy pride of place.

ATHANASII KIRCHERI
Regnum Naturæ
Magne ticum
in triplici Magnete
Inanimato Animato
Sensitivo dispositum.

Arcanis nodis

Indeclinabiliter

Ligantur mutuo

pressa resurget

natura natura delectatur

CAT. 37

# The Subterranean World, *1665*

THE SEISMIC ZONES OF SOUTHERN Italy and Sicily piqued the curiosity of ancient Greek philosophers, Renaissance humanists, and Athanasius Kircher, who made a journey from Rome to Sicily, Malta, Calabria, and Naples in 1637–38. Pietro Bembo's *On Etna* (*De Aetna*), first published in 1494, is a sterling example of the way in which medieval and early Renaissance scholars approached natural phenomena like volcanoes: Bembo's treatise simply provides an anthology of what ancient Greek and Roman writers had to say on the subject. Kircher, some 140 years later, imbued with the spirit of empirical investigation, took the completely opposite approach: he climbed right into the smoking craters of Etna and Vesuvius and in the *Subterranean World* (*Mundus subterraneus*) told his readers exactly what he had seen. Only then did he compare the contributions of other writers, whether ancient or contemporary. The evidence of his own eyes, in other words, counted for as much as any ancient text. Luckily Kircher emerged intact from Etna's crater to tell his tale of geological derring-do. On a similar exploratory trip some 2,100 years before Kircher's, the Greek philosopher Empedocles was said to have fallen in, leaving behind a lone bronze sandal on the volcano's lip. In many ways, he must have been a kindred soul to the German Jesuit: Empedocles' great philosophical poem is called *The Causes of Things*. For Kircher, as for Empedocles, volcanoes provided particularly challenging evidence of the world's complexity. Yet in their apparent chaos, Kircher looked for proof of divine order, and found it:

> And if the first principles of the World are like this, then they must certainly have eternal operations through which they will not fail over the course of time, together with the destruction of all Nature. For this reason that endless duration and perpetuity of the natural operations and Elements comes into being, which cannot be imitated by any human art or industry, for they belong only to the Omnipotence of the divine art. Although its parts may seem to perish, nothing can perish or be lost entirely according to its whole being, so long as it remains subject to the dictates of Nature's law. Hence that perpetual, tireless whirling of the Stars comes into being, ever proceeding by some constant and inviolable law over so many myriads of years; hence the

motions of the Elements is ever invariable amid Nature's great variety. Hence the cycling of the Waters in and around the Earth is perpetual, and as the Wise Man testifies: "*All the rivers run into the sea; yet the sea is not full; unto the place from whence the rivers came, thither they return again* [Eccl. 1:7]; the Sun evaporates the Sea by its vapors, the vapors are dissolved in rain, whence they are extracted, and are soon restored. (Athanasius Kircher, *Mundus subterraneus* [1678], 1.198)

For Athanasius Kircher, nature was the supreme artist. Here he likened the universe to a magnificent building, borrowing both the architectural conceit and the terms he used from the ancient Roman writer Vitruvius, whose *Ten Books on Architecture* (written ca. 25 B.C.E.) he read with care as a philosopher, aesthetician, and archaeologist:

Thus you see that this wonderful work of the divine wisdom has been established so that it lacks neither ordering, nor design, nor management, in which there is nothing that is not superior, harmonious, magnificent, and beyond all admiration: for whether you observe the ornaments of its types of columns, or the splendor of its ceiling coffers, or the solidity of its foundations, you will necessarily be carried away by well-deserved rapture. (Ibid., 1.63)

38. Athanasius Kircher (1602–1680). *Mundus subterraneus in XII libros digestus*. Vol. 1. Amsterdam: Apud Joannem Janssonium à Waesberge & filios, 1678. John Crerar Collection of Rare Books in the History of Science and Medicine

Athanasius Kircher's Earth has no place for a traditional Hell:

Fire-vomiting mountains visible on the external surface of the Earth show well enough that *the Earth is filled with fires*. (1.194)

When the CREATOR OF THE WORLD AND LAWGIVER OF UNIVERSAL NATURE established the Earth . . . he enclosed Fire within its most inward viscera and in its center as the active principle, which would inform and animate all the rest, and Water as the passive principle, joined by such strict laws of friendship that even though they seem to be contrary to one another, one nonetheless cannot live without the other, so that by their reciprocal dealings first they, then all the things that arise from them, may be conserved. (1.198)

39. Pietro Bembo (1470–1547). *De Aetna*. In *Petri Bembi opuscula aliquot, quae sequenti pagella connumerantur*. Lyon: Apud Gryphium, 1532.

Pietro Bembo was a Venetian aristocrat, the son of a powerful, unscrupulous diplomat, Bernardo Bembo. Pietro never shared in his father's political success; he lost every election in which he ran for office in the Republic of Venice. Instead he established himself as a writer, whose elegant style in both Latin and Italian exerted profound effects on the work of his contemporaries. In 1540, at the age of seventy, he was granted a cardinal's hat by his old friend Alessandro Farnese, who had been elected Pope Paul III in 1534. Bembo's treatise on Sicily's Mount Etna, first published in 1494, is typical of his literary work: it is a collection of ancient writers' descriptions of the volcano, more important now for its physical presentation than for its content; indeed Bembo's highly successful career was one long triumph of style over substance. *On Etna* (*De Aetna*) was first published in 1494, with the collaboration of Bernardo Bembo, by the famous Venetian printer Aldus Manutius. The typeface that Aldus designed for this small book drew on ancient Roman inscriptions for its capital letters and modeled its minuscule on the elegant manuscript hand of the Carolingian era (ninth century), which was wrongly thought by fifteenth-century Italian scholars to be the true script of the ancient Romans. Pietro and Bernardo also added a new punctuation mark to the printer's repertory: the semicolon. The modern typeface known as "Bembo" is the direct descendant of the type used in the first edition of *On Etna*. The German-born printer Sebastian Gryphius (1493–1556) was well aware of Aldus' precedent when he published this selection of Bembo's Latin writings for his international clientele.

40. Athanasius Kircher (1602–1680). *Mundus subterraneus, in XII libros digestus.* Vol. 1. Amsterdam: Apud Joannem Janssonium & Elizeum Weyerstraten, 1665.
John Crerar Collection of Rare Books in the History of Science and Medicine

This illustration is based on Kircher's own sketch, still preserved in Rome's National Library (Biblioteca Nazionale Centrale). The accompanying text reads:

> The crater is never without growlings and groanings, which it emits at such horrible volume that they make the Mountain tremble. Let anyone who wants to look upon the wondrous power of GOD GREATEST AND BEST go to one of these mountains, and terrified by the unutterable effects of Nature's miracles, dumbstruck at the same time in their innermost heart be compelled to utter these words: *O the Depth of the riches both of the wisdom and knowledge of GOD! How unsearchable are thy judgments, and thy ways past finding out* by which Thou hast created the World! [Rom. 11:33]. (1.202)

41. Photograph reproduced from Athanasius Kircher (1602–1680). *Mundus subterraneus in XII libros digestus.* Vol. 2. Amsterdam: Janssonio-Waesbergiana, 1678.

Unlike some of his contemporaries, Kircher believed that the images of animals and plants embedded in fossils, like the fish in this image, were the work of human hands:

> And first of all, indeed, with rocks [Nature] performed all the tasks of a good Geometer, when she not only marked out their surfaces with lines, points and every kind of figure, but also adorned their solid bodies with every kind of polygon. Nor did her industry stop here, for rising to higher challenges, she imitated her canopy in the jeweled kingdom of the heavens, which she outfitted in wondrous order with the Sun, Moon, and Stars. Then progressing to Optics, she depicted rivers, forests, meadows, mountains, and seas in many stones according to the most exact standards of perspective, so that the hand of no Optician could ever be seen who wanted to go beyond her industry. Then, embracing the arts of Painting and Sculpture, she drew whatever natural phenomenon came to her attention, and in the variety of many figures she also played with a wonderful kind of sculpture in her work. (2.24)

42. Athanasius Kircher (1602–1680). *Mundus subterraneus in XII libros digestus.* Vol. 1. Amsterdam: Apud Joannem Janssonium à Waesberge & filios, 1678.

Although Kircher was enjoined by the Jesuits' 1599 curriculum to teach the cosmology of Aristotle, he knew that the sun was not a perfect sphere, and said so, despite the fact that Aristotle declared otherwise:

> The Sun is a spherical body not in a mathematical, but in a physical sense, constant in its roughness and unevenness, which might reasonably seem strange had the excellent observations made by the telescope not reassured us. But that it is a fiery, rough, and uneven body has been noted recently by . . . experiment. (Athanasius Kircher, *Ars magna lucis et umbrae* [1646], p. 6)

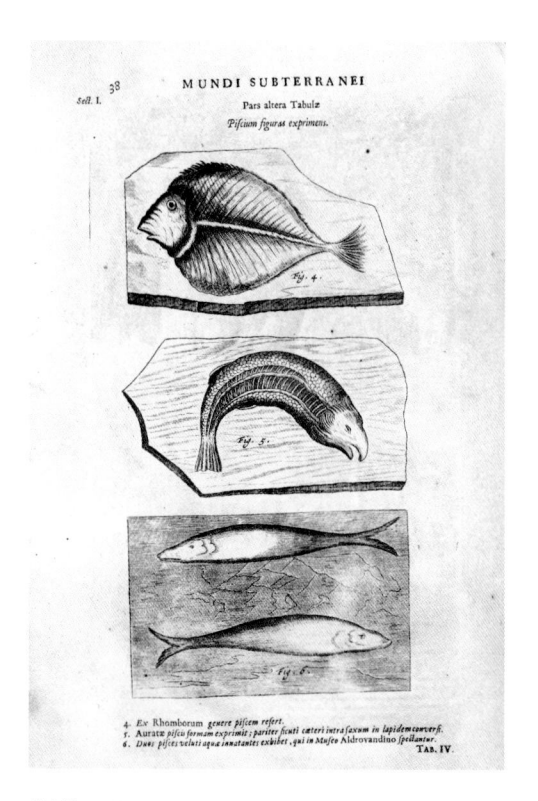

CAT. 41

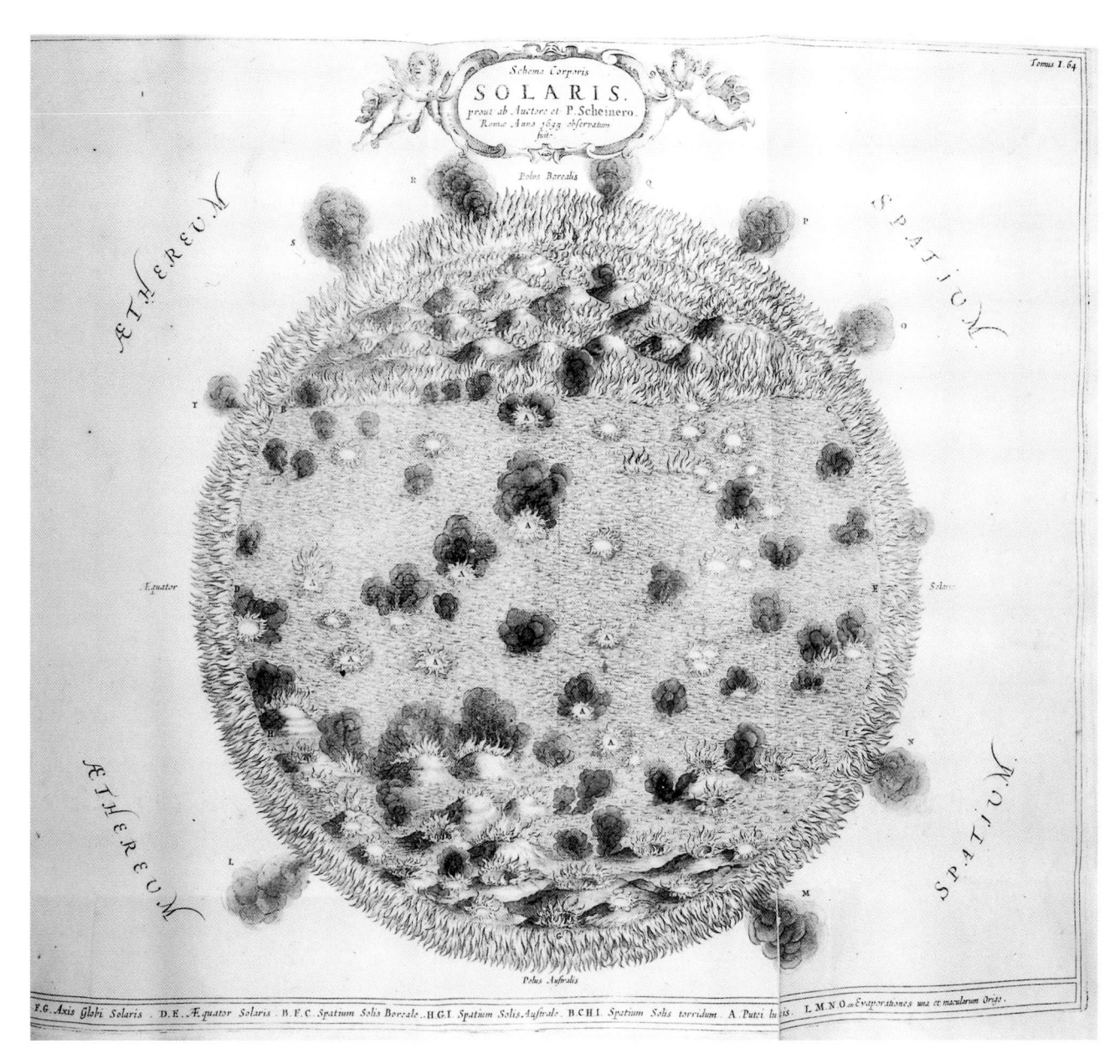

CAT. 42

43. Athanasius Kircher (1602–1680). *Diatribe de prodigiosis crucibus*. Rome: Sumptibus Blasij Deuersin, 1661.

When Vesuvius erupted in 1661, small crosses began to appear on the clothing of terrified Neapolitans. In this little booklet, Kircher showed readers how stains spread in perpendicular patterns across woven linen, and identified the "prodigious crosses" of Naples as tiny particles of volcanic ash, suspended in the atmosphere, that had landed on linen clothing and produced the cross-shaped stains. They were not miracles, in other words, but only the result of the regular workings of nature. On the other hand, Kircher believed in all sincerity that Saint Agatha's intercession could stop Etna from erupting, just as San Gennaro (Saint Januarius) is still reputed to do for Vesuvius.

# *Atlantis*

THE STORY OF ATLANTIS was first told by Plato in his dialogues *Timaeus* and *Critias*. In the fifteenth century, the Florentine philosopher-physician Marsilio Ficino translated these Greek texts for readers of Latin, and the myth of Atlantis was launched in the Renaissance imagination. At the same time, the Age of Exploration pushed European ships across the Atlantic Ocean, the ostensible site of the lost continent. For Athanasius Kircher, the tale of the lost continent of Atlantis furnished proof that even solid ground was part of the dizzying geological change that racked a cosmos created only 6,000 years before. Kircher identified the remnants of Atlantis with the Azores, a tradition that the inhabitants of those Atlantic islands proudly maintain to this day. His younger contemporary, the Padua-educated Swedish polymath Olof Rudbeck (1630–1702), placed Atlantis elsewhere: in his native Sweden. In Rudbeck's patriotic eyes, Plato's account of the great lost city perfectly described the Viking earthworks of Uppsala. His 900-page *Illustrated Atlantis*, simultaneously printed in Latin as *Atlantica illustrata* and in Swedish as *Atland eller Manheim*, took nearly thirty years to publish (1679–1702), but its theories convinced few scholars outside Sweden. Rudbeck's son, Olof, Jr. (1660–1740), an accomplished naturalist, continued his father's antiquarian researches, claiming that Swedish was the closest surviving linguistic relative of the primeval Hebrew spoken by Adam and Eve in the Garden of Eden—which, according to Olof Rudbeck, Sr., had also been set on Swedish soil.

44. Athanasius Kircher (1602–1680). *Mundus subterraneus, in XII libros digestus.* Vol. 1. Amsterdam: Apud Joannem Janssonium & Elizeum Weyerstraten, 1665.
John Crerar Collection of Rare Books in the History of Science and Medicine

For Kircher the fate of Atlantis, shown here, taught an important lesson:

> The Conversions of the Terrestrial Globe are so large and so horrible that they lay bare both the infinite power of GOD and the uncertainty of human fortune, and warn the human inhabitants of this Geocosmos that as they recognize that nothing is perpetual and stable, but that all things are fallible, subject to the varying fates of fortune and mortality, they should raise their thoughts, studies, soul and mind, which can be satisfied by no tangible object, toward the sublime and eternal Good, and long for GOD alone, in whose hands are all the laws of Kingdoms, and the boundaries of universal Nature. (1.83)

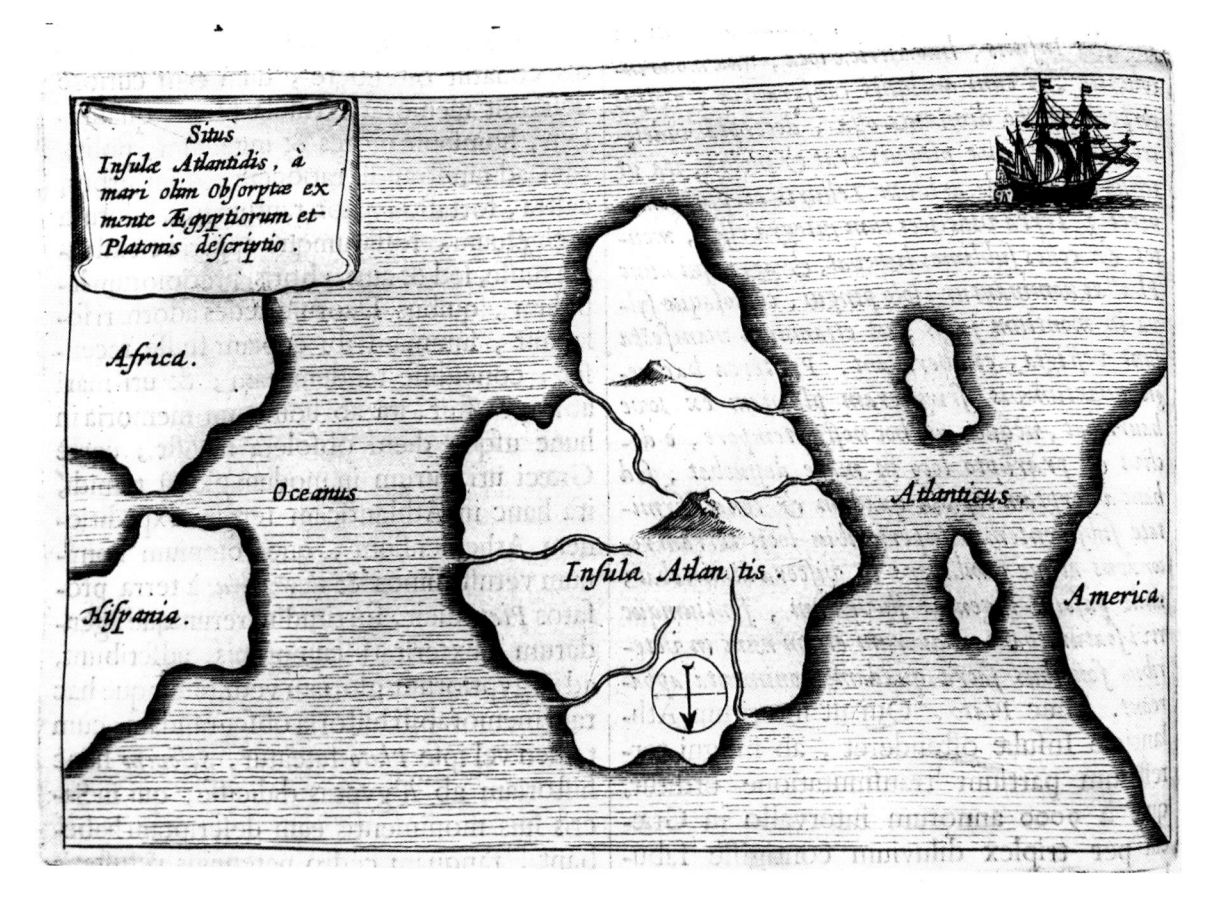

CAT. 44

45. Olof Rudbeck, Jr. (1660–1740). *Atlantica illustrata.* Uppsala: Literis Wernerianis, 1733.

In *The Illustrated Atlantis* (*Atlantica illustrata*), Olof Rudbeck, Jr., followed in his father's footsteps by offering proof that Swedish is the language most closely related to the Hebrew spoken by God, Adam, and Eve.

46. Plato (428/427–348/347 B.C.E.). *Hapanta ta tou Platonos.* [Venice: in aedib. Aldi, et Andreae soceri, 1513].
Helen and Ruth Regenstein Collection of Rare Books

This first printed edition (*editio princeps*) of Plato's dialogues *Timaeus* and *Critias* was issued by Aldus Manutius in 1513. Its elaborate script, very difficult for modern classicists to read without practice, mimics the Byzantine cursive script of editor Marcus Musurus.

47. Plato (428/427–348/347 B.C.E.). *Opera.* Translated by Marsilio Ficino. [Venice: per Bernardinu de Choris & Simone de luero impēsis Andree Toresani de Asula, 1491].
Gift of Harry Wohlmuth

In this very early edition of Ficino's translations into Latin from Plato's Greek, marginal notes emphasize "the Pillars of Hercules," "huge earthquakes," and "a one-day flood."

> [There was] an island before the straits that you call, as you say, the Pillars of Hercules. Now the island was bigger than Libya and Asia put together.... On this island of Atlantis there was a great and marvelous confederation of kings that governed all the island, many other islands and parts of the continent. Within [the straits] they governed Libya as far as Egypt and Europe as far as Etruria.... In a later time, huge earthquakes and floods occurred, and in one terrible day and night... the island of Atlantis disappeared, sinking beneath the sea. (*Timaeus*, 25B–D)

# Galileo's Predecessors and Rivals

ALTHOUGH THE GREAT PHILOSOPHER ARISTOTLE argued that the heavenly bodies were perfect, unblemished spheres, his opinions were not universally accepted by ancient thinkers. In the heyday of the Roman Empire, Plutarch of Chaeronea (ca. 47–120 C.E.) presented various explanations for the moon's mottled appearance in his essay *On the Face that Appears on the Sphere of the Moon* (*De facie quae in orbe lunae apparet*), ultimately opting for a moon with mountains and valleys like the earth's. But it was the invention of the telescope that revealed the moon's irregularity beyond any doubt, as well as the existence of sunspots, the presence of satellites around the planets, and an infinitely great number of stars. Aristotle's opinions could no longer hold up to the weight of empirical evidence, and Athanasius Kircher was one of those willing to say so, if only through the mouth of his guardian angel, Cosmiel:

> *Cosmiel*: You are mistaken, and greatly so, if you persuade yourself that Aristotle has entirely told the truth about the nature of the supreme bodies. It is impossible that the philosophers, who insist upon their ideas alone and repudiate experiments, can conclude anything about the natural constitution of the solid world, for we [angels] observe that human thoughts, unless they are based on experiments, often wander as far from the truth as the earth is distant from the moon. (Athanasius Kircher, *Iter exstaticum* [1660], p. 97)

Kircher was also an early believer in what he called the earth's "Athmo-sphaera," the shroud of air that kept living things from suffocating in the thin air of outer space. Many observers also believed that the moon must have a similar feature, including Kircher's student Gaspar Schott. Based in Germany, Schott read the prohibited books of Galileo and Bruno with impunity. Schott noted in his revision of the *Ecstatic Heavenly Journey* (*Iter exstaticum*):

> Many people, since the recent use of the Optical Tube, now accept that around [the moon] there is either air, or a denser aether; so Kepler, Maestlin, Galileo, Longimontanus, Giordano Bruno, Antonio Maria Reita, and Scheiner. (Ibid., p. 69)

48. Plutarch (ca. 47–120). *De facie*, from *Moralia*. In *Plutarchi Chæronensis omnium, quae extant, operum*. Vol. 2. [Frankfurt: Andreae Wecheli, Claudium Marnium, & Ioannem Aubrium, 1599].

Before writing his *Parallel Lives of the Great Greeks and Romans*, Plutarch of Chaeronea composed a large number of short essays, later collected under the title *Moral Writings* (*Moralia*). His essay *On the Face that Appears on the Sphere of the Moon* (*De facie quae in orbe lunae apparet*) is partly based on Plato's *Timaeus*, with its story of Atlantis, and features a traveler who comes from the continent on the other side of the Atlantic Ocean. Johannes Kepler believed on the basis of this account that Plutarch must have had knowledge of the New World.

> Behold the moon as she is in herself: her magnitude and beauty and nature, which is not simple and unmixed but a blend as it were of star and earth . . . just as our earth contains gulfs that are deep and extensive . . . so those features [of the so-called face] are depths and hollows of the moon. (*De facie*, 943E–944B; translated by Harold Cherniss and William C. Helmbold, *Plutarch's Moralia* [London: William Heinemann, 1927], vol. 12)

49. Christopher Clavius (1538–1612). *In sphæram Iohannis de Sacro Bosco commentarius*. Venice: Apud Io. Baptistam Ciottum, 1601.

Christoph Clavius held the chair of mathematics at the Jesuits' Collegio Romano from 1564 until his death in 1612. He was a scholar of immense distinction, as well as a compelling teacher. When the order instituted a new curriculum in 1599 that required professors to teach the cosmology of Aristotle, it must have come as a terrible blow to Clavius, who was the first person to invite Galileo to lecture in Rome. His astronomical commentary on a work by the medieval English astronomer John Holywood (Italianized as Sacrobosco; fl. 1230) was immensely influential in his day; the University of Chicago Library holds several editions, beginning with the 1581 first edition, written before the revised Jesuit curriculum had been put into effect.

This third edition of Clavius was published in Venice by Giovanni Battista Ciotti, the prominent printer and bookseller, an acquaintance of Paolo Sarpi and Giordano Bruno who was twice examined by the Venetian Inquisition after Bruno's arrest in 1592. It is probably no accident that Ciotti's symbol was the sun, with implications for his own thinking about cosmology. Furthermore,

CAT. 48

we know from Kircher's testimony, and by Clavius' behavior, that the great mathematician's sentiments lay with Copernicus and Galileo:

> Nonetheless the good Father Athanasius, whom we saw passing through here quickly, could not restrain himself from informing us . . . that Father Malapertius and Father Clavius themselves in no way disapprove of Copernicus' opinion. Indeed,

they would hardly differ from it at all if they were not compelled and obliged to write according to the common suppositions of Aristotle, which Father Scheiner himself does not endorse except by force and obedience, even if he has no trouble admitting not only mountains on the body of the moon, but also valleys and seas on its surface, even trees and plants and animals. (Nicolas Claude Fabri de Peiresc to Pierre Gassendi, August 27, 1633; from Philippe Tamezey de Larroque, ed., *Lettres de Peiresc*, vol. 4, *Lettres 1626–1637* [Paris: Imprimerie Nationale, 1893], p. 354)

50. Orazio Grassi (1583–1654). *Ratio ponderum libræ et simbellæ*. Naples: Excudebat Matthæus Nuccius, 1627.
Berlin Collection

Orazio Grassi was professor of mathematics at the Collegio Romano from 1617 to 1624, and again from 1626 to 1628 or 1632. He began debating Galileo's work on comets in a pseudonymous pamphlet of 1619, *The Astronomical and Philosophical Balance* (*Libra astronomica ac philosophica qua Galilaei Galilaei Opiniones de cometis a Mario Guiducio in Florentina Academia expositae*) (Perugia, 1619). Galileo's reponse was his scintillating *Assayer* (*Il saggiatore*), to which this essay in turn served as a reply. Despite the fact that the page on display chides Galileo for weak argumentation, the overall tone is respectfully conciliatory; Grassi and Galileo both loved the give-and-take of a good controversy. From a modern standpoint, Grassi's interpretation of comets was more nearly correct than Galileo's. Contemporaries, however, thought Galileo the better debater, and this book of Grassi's exerted comparatively little influence on the course of scholarly thought. That may not in fact have been its purpose—it may have been a kind of peace offering, albeit one with a kick.

51. Antonio Reita (1597–1660). *Nouem stellæ circa Iouem, circa Saturnum sex, circa Martem non-nullæ*. Louvain: Andree Bouvetij, 1643.
John Crerar Collection of Rare Books in the History of Science and Medicine

This compilation was assembled by the prolific Spanish Cistercian Juan Caramuel y Lobkowitz (1606–1682). It begins with a letter from Father Antonio Reita, an eager Capuchin astronomer working in Cologne, who claimed to have seen five new satellites of Jupiter in addition to Galileo's four "Medicean Stars," as well as six around Saturn and several around Mars. The French royal astronomer Pierre Gassendi cast doubt on Reita's report by suggesting that the five new stars were part of the constellation Aquarius. In the volume's longest essay, Lobkowitz—who was already embarked on an illustrious international career as diplomat, abbot, and author—invited his similarly influential dedicatee, the young prelate (and future pope) Fabio Chigi, to adjudicate between the two extreme positions. Lobkowitz gently suggested that Reita may have seen reflections from one of his telescope's multiple lenses, but he insisted that such enthusiasm for observational astronomy should by all means be encouraged.

# Copernicus and Copernicanism

THE POLISH PHYSICIAN NICOLAUS COPERNICUS (1473–1543) had arrived at his concept of an earth that moved on its axis and circled the sun with other planets by about 1530. Ideas from the work circulated in manuscript before it was published, at the behest of Georg Joachim Rheticus, the year of its author's death. It took over a century for the Copernican world view to find its full audience and influence—although Galileo and Kepler were supporters, its challenge to Church-held beliefs forced Kircher and others to mute their advocacy:

> This system is called Copernican after Nicolaus Copernicus the Pole, who finally completed what had once been partly devised by Philolaus the Pythagorean, and Aristarchus the Samian, and then resuscitated by Nicolaus Cusanus, and he supported it with many arguments and ingenious hypotheses; afterward, almost all the non-Catholic mathematicians have followed him and some among the Catholics, for whom, not surprisingly, their talents and their pens itch to report something new. (Athanasius Kircher, *Iter exstaticum* [1660], p. 38)

52. Nicolaus Copernicus (1473–1543). *De reuolutionibus orbium cœlestium*. Nuremberg: apud Ioh. Petreium, 1543.

Although Copernicus is now understood to have revolutionized our understanding of the cosmos, there is little hint of the profound influence that *On the Revolutions of the Heavenly Spheres* (*De revolutionibus orbium coelestium*) would have, or the religious controversy it would cause, on the title page of this first edition:

> Six books on the revolutions of the heavenly spheres, by Nicolaus Copernicus. You have in this recently conceived and published work, diligent reader, the motion of the stars, both the fixed stars and the planets, restored according to the observations of the ancients and the moderns, and adorned besides with new and admirable hypotheses. You also have convenient tables, by which you may easily calculate the same for any time you like. Therefore buy, read, and enjoy!

The famous diagram of the solar system shown here illustrates the radical nature of Copernican theory, with the sun at the center and the concentric orbits of the planets, including the earth.

53. Galileo Galilei (1564–1642). *Istoria e dimostrazioni intorno alle macchie solari*. Rome: Appresso Giacomo Mascardi, 1613.

A talented musician and brilliant scientist, Galileo was also a draftsman of distinction. The plates for this work on sunspots, taken directly from his drawings, show that he was not only a sharp-eyed observer, but also unusually able to record those observations in an immediately intelligible way.

54. John Wilkins (1614–1672). *The discovery of a world in the moone*. London: Printed by E. G. for Michael Sparke and Edward Forrest, 1638.

In this essay, published in Protestant England, John Wilkins, Bishop of Chester, suggested that "'tis probable there may be another habitable world in the moone." Wilkins mentioned the Copernican system in passing, commenting that: "how horrid soever this may seeme at the first, yet it is likely to be true." Wilkins openly referred to his readings from the heretic philosopher Giordano Bruno. Bruno's thought had special resonance in

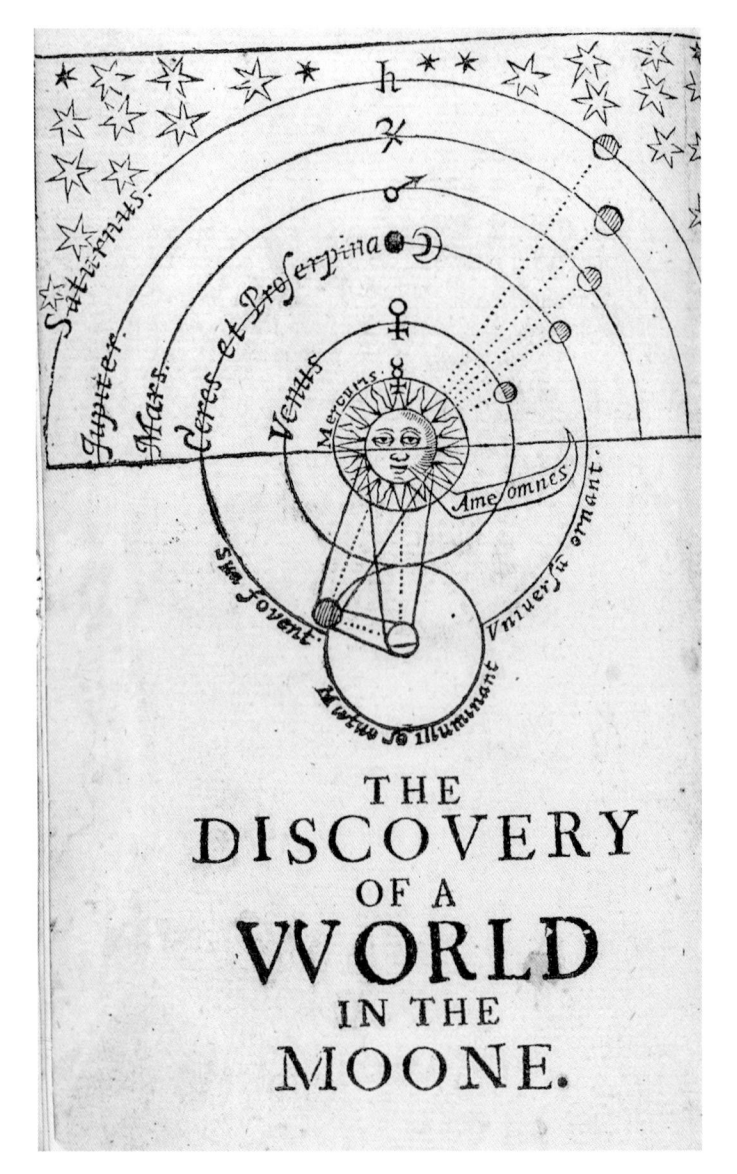

England, where he had lived, mostly in London, from 1583 to 1585. Despite the Italian philosopher's tendency to fly into incandescent rages—as he did during a lecture at Oxford in 1584 and seems to have done in private as well—he exerted considerable influence on some prominent members of the Elizabethan court, and wrote his most brilliant dialogues, all six composed in Italian vernacular, under the beneficent influence of the queen he called "the one and only Diana."

55. John Wilkins (1614–1672). *A discourse concerning a new world & another planet.* London: Printed by Iohn Norton for Iohn Maynard, 1640.

After publishing his *Discovery of a world in the moone,* John Wilkins, Bishop of Chester, found that he had an enthusiastic reading public. This edition of 1640, the third, has increased dramatically in size from its predecessors. In addition, Wilkins revealed himself unequivocally as a Copernican: the "new planet" to which he refers in his title is the planet Earth. This volume is open to a page on which Wilkins cited two of the early defenders not only of a sun-centered planetary system, but also, far more radically, of an infinite universe: the fifteenth-century German cardinal Nicholas of Cusa and the late sixteenth-century Italian philosopher Giordano Bruno. Wilkins himself became one of an illustrious group of scientists who met in London at Gresham College, an institution for adult education that had been founded in 1597 but gained real momentum in the 1640s because of his own galvanizing presence.

CAT. 54

# Tycho Brahe's Observatory and Cosmological System

THE MOST GIFTED OBSERVATIONAL ASTRONOMER of the late fifteenth century was the Danish aristocrat Tycho Brahe (1546–1601), who built his remarkably equipped observatories on his feudal domain, the island of Hven in the Danish Sound. Using his famously accurate records of the movements of stars and planets, he devised a cosmological system that struck a compromise between the Copernican heliocentric system and the traditional earth-centered universe: to his mind, the planets revolved around the sun, while the sun and the sphere of the fixed stars revolved around the earth. This was the system to which Athanasius Kircher and many of his associates adhered in public; their private thoughts were another matter.

56. Tycho Brahe (1546–1601). *Astronomiæ instauratæ mechanica*. Nuremberg: apud Levinum Hulsium, 1602.

Although he served as unofficial court astronomer to Frederick II of Denmark for twenty years, Tycho Brahe, regarded as the true founder of observational astronomy, was a feudal lord by profession; the king granted him the island of Hven in the Danish Sound as his private fiefdom in 1576. There Tycho outfitted two observatories: Uraniborg, "Heavenly Castle," and the underground Stjerneborg, "Starry Castle." This engraving of Tycho at work in the luxuriously appointed interior of Uraniborg copies a portrait fresco from the observatory's wall. The engraved image, like the original fresco itself, is framed by the immense instrument called the "Tychonian quadrant," or wall-quadrant, which occupied most of the wall's surface. Tycho's working uniform is elaborate, befitting a peer of the realm. In 1597, however, Tycho fell from favor and sought another source of patronage. In the same year, he found a suitable sponsor in Rudolph II, King of Bohemia, and settled in Prague. In 1600 Johannes Kepler joined Rudolph's staff, and was reportedly the person who heard Tycho's last words, "May I not have lived in vain."

57. Photograph reproduced from Tycho Brahe (1546–1601). *Astronomiæ instauratæ mechanica*. Nuremberg: apud Levinum Hulsium, 1602.

This is an exterior view of Uraniborg, Tycho Brahe's remarkable observatory, set within an elaborate garden on his private island of Hven. Here his sister Sofie studied alchemy, astrology, and medicine, becoming a successful professional woman who died at the age of 84. Tycho also deployed a small army of assistants to make his instruments, help him with his records, and publish his books. Tycho's superb equipment and his own meticulous care gained his astronomical observations nearly universal admiration; characteristically, however, Galileo, who was not above petty jealousies, loathed his Danish rival. Unfortunately, both Uraniborg and Stjerneborg were destroyed not long after Tycho's death in 1601.

58. Photograph reproduced from Tycho Brahe (1546–1601). *Astronomiæ instauratæ mechanica.* Nuremberg: apud Levinum Hulsium, 1602.

The name of Tycho's underground observatory, Stjerneborg ("Starry Castle"), has been Latinized in this engraving as "Stellaeburg." This inscription stood on its Ionic portal:

> Consecrated to the all-good, great God and Posterity. Tycho Brahe, Son of Otto, who realized that Astronomy, the oldest and most distinguished of all sciences, had indeed been studied for a long time and to a great extent, but still had not obtained sufficient firmness or had been purified of errors, in order to reform it and raise it to perfection, invented and with incredible labour, industry, and expenditure constructed various exact instruments suitable for all kinds of observations of the celestial bodies, and placed them partly in the neighbouring castle of Uraniborg, which was built for the same purpose, partly in these subterranean rooms for a more constant and useful application, and recommending, hallowing, and consecrating this very rare and costly treasure to you, you glorious Posterity, who will live for ever and ever, he, who has both begun and finished everything on this island, after erecting this monument, beseeches and adjures you that in honour of the eternal God, creator of the wonderful clockwork of the heavens, and for the propagation of the divine science and for the celebrity of the fatherland, you will constantly preserve it and not let it decay with old age or any other injury or be removed to any other place or in anyway be molested, if for no other reason, at any rate out of reverence to the creator's eye, which watches over the universe. Greetings to you who read this and act accordingly. Farewell! (Translation from Tycho Brahes Glada Vänner, Lund, Sweden. http://www.nada.kth.se/~fred/tycho/tychoquote.html; searched January 11, 2000)

59. Pierre Gassendi (1592–1655). *Tychonis Brahei, equitis dani, astronomorum coryphæi vita.* Paris: Apud viduam Mathurini Dupuis, 1654.
John Crerar Collection of Rare Books in the History of Science and Medicine

Pierre Gassendi wrote his biography of Tycho Brahe during the remarkably productive years of his retirement in Provence. The engraver of this portrait dealt discreetly with a delicate problem: Tycho lost his nose as a student at Rostock; it was sliced off in a rapier duel with another

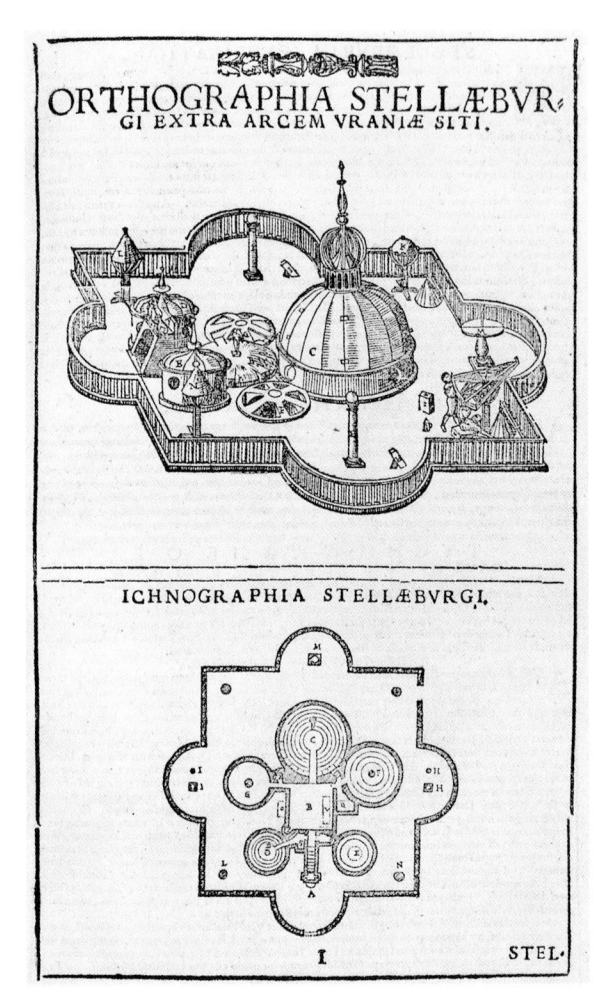

CAT. 58

Danish nobleman, Manderup Parsberg. The elegant Dane thereafter made do with an ivory prosthesis. A faint dotted line on the engraving marks where flesh ends and ivory begins.

60. Photograph reproduced from Pierre Gassendi (1592–1655). *Tychonis Brahei, equitis dani, astronomorum coryphæi vita.* Paris: Apud viduam Mathurini Dupuis, 1654.

Tycho's observations of celestial motion convinced him that the planets must revolve around the sun, but he continued to believe that the sun and stars revolved around the earth. This image of the Tychonic cosmological system illustrates his theory.

| | |
|---|---|
| RUDER | BRAHE BILLER |
| | ULFSTANDER |
| LONGER | RONNOR |
| ROSENKRANS | TROLLER |
| AXELLSØNNER | LONGER |
| MARCKEMAN | ROSENSPAR |
| KABBELER | STORMVASE |
| GULDENSTEREN | AXELLSØNER |

NON HABERI — SED ESSE

EFFIGIES TYCHONIS BRAHE OTTONIDIS DANI DÑI
DE KNVDSTRVP ET ARCIS VRANIENBVRG IN INSVLA
HELLISPONTI DANICI HVENNA FVNDATORIS INSTRV-
MENTORVMQVE ASTRONOMICORVM IN EADEM DISPO-
SITORVM INVENTORIS ET STRVCTORIS ÆTATIS
SVÆ ANNO 40. ANNO DÑI 1586 COMPL.

# *Galileo's* Starry Messenger *and the Infinite Universe*

In 1610 Galileo Galilei, professor of mathematics at the University of Padua, turned his new telescope, with its thirtyfold power of magnification, toward Jupiter, and saw satellites moving about it. In a book with a beguilingly sprightly title, *The Starry Messenger* (*Sidereus nuncius*), he named these newly discovered bodies the "Medicean stars" to curry the favor of Cosimo II de' Medici, grand duke of Tuscany, identifying each satellite with one of the grand duke's sons. Galileo was rewarded almost immediately by an appointment as chief mathematician and philosopher to the grand duke, and chief mathematician at the University of Pisa, an appointment that did not require him ever to appear in his native city.

The idea of an infinite universe extended back to ancient times; Plutarch seemed to favor it in his essay *On the Face that Appears on the Sphere of the Moon*: "After all, in what sense is earth situated in the middle, and in the middle of what? The sum of things is infinite, and the infinite, having neither beginning nor limit, cannot properly have a middle" (925F). In the fifteenth century, the German cardinal Nicholas of Cusa (Nicolaus Cusanus) revived the idea, exerting tremendous influence over the sixteenth-century Italian philosopher Giordano Bruno. Both Galileo and Kircher probably tended toward the same opinion, although they were careful not to say so outright; Bruno, after all, had burned at the stake for his ideas in 1600.

From the transcripts of the trial of Giordano Bruno before the Inquisition of Venice (Luigi Firpo, *Il processo di Giordano Bruno* [Rome: Salerno Editrice, 1993]):

> He says that the stars are also angels in the following words: 'And the stars are angels, rational bodies with souls and while they praise God and proclaim His power and might, by means of those lights and writings sculpted in the firmament *The Heavens declare the glory of God*. Angels means nothing other than messengers and interpreters of the divine voice and of nature, and these are perceptible and visible angels, aside from those other invisible and imperceptible ones.

> In his conversations, [Bruno] asserted that there were many worlds, and that the world was a star, and that this world appeared as a star to the other worlds, just as the other worlds shine down on us as stars.

One evening he beckoned Francesco Napolitano to the window and showed him a star, saying that it was a world and that all the stars were worlds. (P. 303)

He said that God needed the world as much as the world needed God, and that God would be nothing without the world, and for this reason God did nothing but create new worlds. (P. 268)

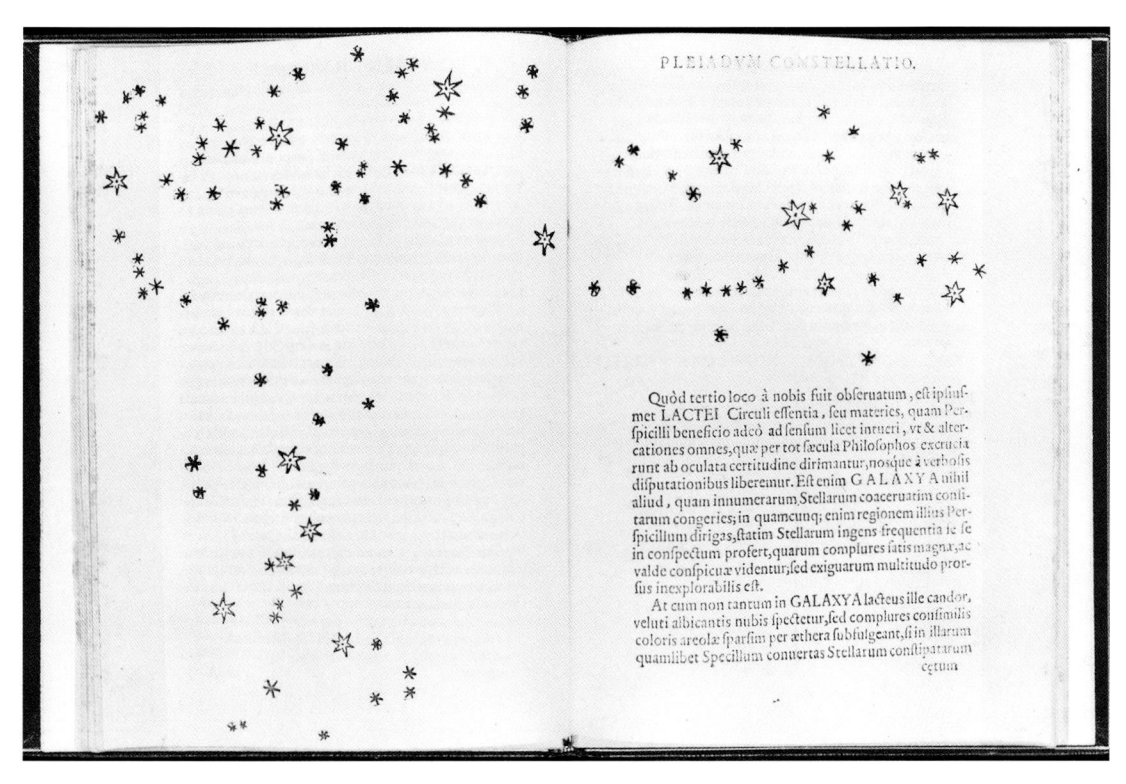

CAT. 61

61. Galileo Galilei (1564–1642). *Sidereus nuncius*. Venice: Apud Thomam Baglionum, 1610.

The woodcut image in the *Starry Messenger* (*Sidereus nuncius*) of the constellation of the Pleiades, expanding as it does beyond the margins of the text block, presents a strikingly apt image of what Galileo's observations with the telescope were doing to conventional concepts of the universe: bursting their boundaries by disclosing a sky filled with an unanticipated profusion of beautiful, but apparently randomly distributed, stars. As the telescope revealed these new legions of heavenly bodies, Galileo and Kepler both held stubbornly to the conviction that they were contained within a finite universe; in fact, Kepler's reply to Galileo's *Starry Messenger*, displayed in this case, breathes an audible sigh of relief that Galileo had found moons revolving around Jupiter rather than one of the "fixed stars." Had a fixed star been found with satellites, as Kepler says, they might have had to contemplate Giordano Bruno's idea of an infinite universe!

> In the first place, I rejoice that you have restored me not a little by your labors. If you had found planets circling one of the fixed stars, there among Bruno's infinities I had already prepared my prison shackles, that is, my exile in that Infinity. Thus you freed me from the great fear that I had conceived when I first heard about your book . . . because you say that these four Planets run their course around Jupiter rather than one of the fixed stars. (Johannes Kepler, *Dissertatio cum nuncio sidereo* [1610], P. 9 verso)

62. Johannes Kepler (1571–1630). *Dissertatio cum nuncio sidereo*. Florence: Apud Io. Antonium Canêum, 1610.
From the Library of Joseph Halle Schaffner

Galileo sent a copy of his *Starry Messenger* to Johannes Kepler in Prague, to the latter's delight. Kepler read the book immediately, and drafted this complimentary reply in a long letter. Now it was Galileo's turn to be elated. He handed Kepler's letter to a Florentine printer for immediate publication. In this response to Galileo's book, Kepler revealed his initial fears that Galileo's experimental evidence would force him to abandon his belief that the universe was a finite set of spheres separated by musical intervals, and his relief that this was not the case. Galileo's own opinion about the structure of the universe probably corresponded more closely to Giordano Bruno's infinite space, although he never said so openly. Nonetheless Kepler recognized the impact of Bruno's thought on Galileo, and gently chided him here for failing to mention that influence. Kepler, of course, had far less to fear from the Inquisition; Prague may have been Catholic, but it enjoyed much greater intellectual freedom than Italy.

> For the glory of this world's Architect greatly exceeds that of the contemplator of such glory, however ingenious. The former, after all, drew the principles of its creation from within Himself, whereas the latter, after great effort, scarcely recognizes the expression of such principles in that same creation. Certainly those who can conceive the causes of phenomena in their minds before the phenomena themselves have been revealed are more like Architects than the rest of us, who consider causes only after they have seen the phenomena. Do not, therefore, Galileo, begrudge our predecessors their proper credit . . . you refine a doctrine borrowed from Bruno. . . . (P. 10 recto)

63. Giordano Bruno (1548–1600). *De immenso et innumerabilibus*. In *De monade numero et figura liber consequens quinque de minimo, magno & mensura*. Frankfurt: Apud Ioan. Wechelum & Petrum Fischerum, 1591.
Berlin Collection

Bruno published his *On the Immeasurable and Innumerable* (*De immenso et innumerabilibus*) in 1591 as one of a trio of philosophical poems that also includes *On the Minimum* (*De minimo*). Whereas *On the Minimum* explains the most minute particles in the universe in terms of atomic theory, *On the Immeasurable*, shown here, presents Bruno's ideas of an infinite universe composed of an infinite number of systems of fiery stars orbited by colder planets. Despite the influence his ideas were to wield on some of his contemporaries, and the uncanny way in which some of his theories anticipate modern thinking about the cosmos, Bruno disdained empirical observation; to his mind, it was more important to develop hypotheses. Kepler, at least, agreed with him.

> However many suns it is possible to believe in,
> We find a number of planets circling around every one.
> Not because one number wants to exceed another,
> For the suns are numberless suns, the planets numberless planets,
> So numberless that units equal numberless pairs and triads;
> No one dares to say that the cubits in measureless space
> Outstrip the number of paces, or of parasangs.
> Ask not for a finite number, or for finite numbers.
> Here where there is no place either for numbers or limits
> Number cannot be assigned.
> (PP. 360–61)

64. Athanasius Kircher (1602–1680). *Magnes, siue De arte magnetica*. Cologne: Apud Iodocvm Kalcoven, 1643.
John Crerar Collection of Rare Books in the History of Science and Medicine

In *The Magnet* (*Magnes*), his general treatise on magnetism, Athanasius Kircher refuted Kepler's elliptical orbits, and provided his readers with an image of a circular solar system—hardly a stern refutation of the solar system per se! In Kircher's cosmology, magnetism plays the part that gravity will for Newton; it is the attractive force that binds the universe together. For Kircher this cosmic bonding is the most basic physical expression of the idea that God is Love; even the rocks and the planets love one another. Giordano Bruno expressed the same conviction more radically: for him the entire universe was infused with a great, loving world-soul that extended to infinity, an emanation of the infinite power of divinity.

# The Ecstatic Heavenly Journey

KIRCHER'S TOUR OF THE COSMOS, *The Ecstatic Heavenly Journey*—cast as a dream dialogue between his own alter ego, Theodidactus ("Taught by God"), and his guardian angel, Cosmiel—was first published in Rome in 1655, to immediate criticism by Jesuit censors for its doctrinal irregularities. These readers found Kircher's defense of traditional cosmology halfhearted at best. By contrast, his explanation of the Copernican system, as expressed by the talkative Cosmiel, is clear and concise. Cosmiel also has sharp words about Aristotle, and some ideas about cosmic chemistry that, to the censors at least, verged on outright heresy. Kircher's student Gaspar Schott immediately undertook revision of the text, and published an annotated version in Rome and in Germany, where it reached a wide audience and was reprinted several times. The dialogue must have enjoyed the tacit protection of the reigning pope, Alexander VII, an old friend of Kircher's with a long history of Jesuit ties and strong ideas of his own, for Schott's revisions cite forbidden authors, and nothing was done to alter the content of Kircher's text.

In his introduction to the *Ecstatic Heavenly Journey*, Kircher prepared his readers for a work of science fiction:

> The point and the single intent of this work, once I began it, was to follow a method in this little book to which Hermes Trismegistus, Plato and Lucian (among ancient authors) as well as many Poets and Orators in succeeding centuries have subscribed by laudable custom, a method, I say, that complements a kind of rhetoric that is charming, pleasant, and accommodated to the reader's taste, just as outline and shadow gently temper light and color. Here, I would not actually want to undergo any mystic initiation, any rapture, any revelation of Divinity, no angelic epiphany, no inspiration by the Delphic Oracle—rather, for the sake of clear demonstration and to have the reader more easily assimilate information about hitherto unknown subjects I would like for you to be persuaded that they are exhibited under the wraps of an ingenious fiction, in the guise of a fictitious rapture. (Athanasius Kircher, *Iter exstaticum* [1660], p. 16)

ITER EXSTATICUM
KIRCHERIANUM,
Prælusionibus & Scholijs
illustratum, schematibus
exornatum
à
P. GASP. SCHOTTO
Societatis JESU.
1660.

65. Athanasius Kircher (1602–1680). *Iter extaticum coeleste*. Würzburg: Sumptibus Joh. Andr. & Wolffg. Jun. Endterorum hæredibus, 1660.
John Crerar Collection of Rare Books in the History of Science and Medicine

This is how Gaspar Schott described the genesis of Kircher's book in his annotations to its second edition:

> [Kircher] said he was not about to [write about cosmology], lest he be perceived as an Innovator among the usual Philosophers and Astronomers whose opinions had already been received. . . . He refused again, lest he distract his mind from the *Oedipus*, and then from the *Mundus subterraneus*, which he had begun to write. Finally, one day after he had discoursed at length and brilliantly, as was his wont, on this same subject, and I had repeated my entreaties and desires more insistently; the next day, about noon, he addressed me as follows: "Father, last night I dreamt a remarkable dream. I saw myself led by my guardian angel to the Moon, to the Sun, to Venus, to the rest of the Planets, to the very fixed stars and the outermost boundaries of the universe, and furthermore I found everything that I have so often spoken about to Your Reverence." "So now it is time," I said, "that Your Reverence stop delaying and write down your dream." (P. \*\* 2 recto)

66. Peter Apian (1495–1552). *Astronomicum Cæsareum*. [Ingolstadt, 1540]. Inscribed by Tycho Brahe.
From the Library of Joseph Halle Schaffner

This luxurious collection of astronomical tables has movable paper dials (volvelles) and hand-colored plates. Because it was issued three years before the publication of Copernicus' *On the Revolutions of the Heavenly Spheres*, it presents only an earth-centered cosmos, filled with fantastic beasts. It is important nonetheless for its observations of comets.

67. Photograph reproduced from Peter Apian (1495–1552). *Astronomicum Cæsareum*. [Ingolstadt, 1540]. Inscribed by Tycho Brahe.
From the Library of Joseph Halle Schaffner

The University of Chicago Library's copy of Apian's *Astronomicum* was presented by Tycho Brahe to his student Paul Wittich on October 29, 1580. The bold, assertive handwriting of Tycho's autograph inscription says a good deal about the way the meticulous scientist inhabited a genial, expansive, and self-consciously aristocratic personality—in his penmanship, the lord wins out over the perfectionist.

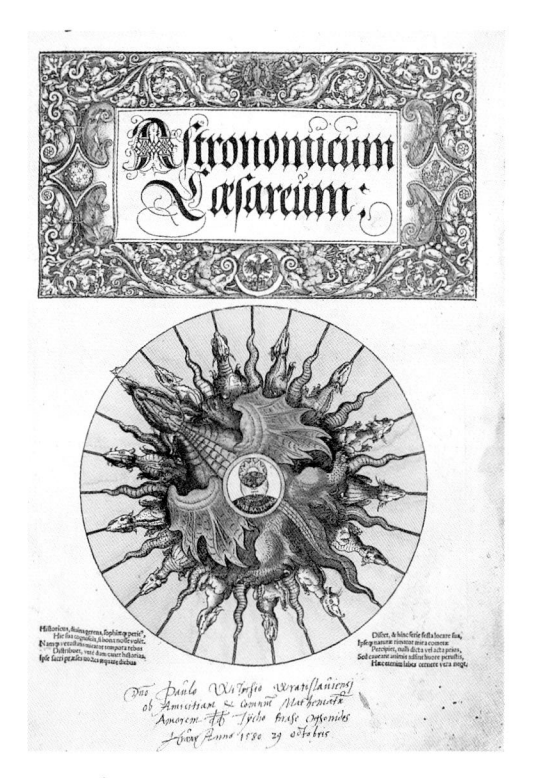

CAT. 67

68. Athanasius Kircher (1602–1680). *Iter exstaticum coeleste*. Würzburg: Sumptibus Johannis Andreæ Endteri, & Wolfgangi Junioris hæredum, 1671.
Presented in Memory of Cora B. Perrine

As a concession to the limits placed on him as a Catholic, Kircher presented diagrams of all the prevailing cosmological theories and claims to subscribe to the modified Tychonic system in which the sun alone circles around the earth. There is little question, however, that the author's real beliefs lay elsewhere, with the Copernican system he was forbidden to adopt. Quoting the love poetry of the biblical *Song of Songs*, Kircher made it clear in this dreamy dialogue that his cosmological researches, like all his other endeavors, are inspired by the desire to know and understand God:

> Excited by these things, I said in my heart: *I will rise now, and go about the* heavenly *city in the streets, and in the broad ways I will seek him whom my soul loveth*: so that when I have found him in his works, *I will hold him and will not let him go* [Song of Songs 3:2, 3:4]. But—o miserable condition of our befogged intellects!—the more I progressed, I discovered, the more I fell short of attaining the glory of the Lord's Majesty. (P. 62)

# *Galileo's* Discourse on Floating Bodies; *Kircher's* Tower of Babel *and* Noah's Ark

IN HIS TREATISE *The Tower of Babel*, Athanasius Kircher addressed an issue that was becoming more and more important in his day: technology. The biblical story of a tower high enough to reach the moon allowed him to discuss the problems of human ingenuity and its limits:

*And the whole earth was of one language, and of one speech.*

*And it came to pass, as they journeyed from the east, that they found a plain in the land of Shinar; and they dwelt there.*

*And they said one to another, Go to, let us make brick, and burn them thoroughly. And they had brick for stone, and slime had they for mortar.*

*And they said, Go to, let us build us a city and a tower, whose top may reach unto heaven; and let us make us a name, lest we be scattered abroad upon the face of the whole earth.*

*And the LORD came down to see the city and the tower, which the children of men builded.*

*And the LORD said, Behold, the people is one, and they have all one language; and this they begin to do: and now nothing will be restrained from them, which they have imagined to do.*

*Go to, let us go down, and there confound their language, that they may not understand one another's speech.*

*So the LORD scattered them abroad from thence upon the face of all the earth: and they left off to build the city.*

*Therefore is the name of it called Babel; because the LORD did there confound the language of all the earth: and from thence did the LORD scatter them abroad upon the face of all the earth.*

(Gen. 11:1–9)

69. Athanasius Kircher (1602–1680). *Turris Babel*. Amsterdam: Ex officina Janssonio-Waesbergiana, 1679.

Lieven Cruyl's engraving of the Tower of Babel is one of the finest images ever executed for Kircher's books. Kircher, who so evidently loved ingenious machines, had no doubt that inventions conceived in a charitable spirit were all to the good. As for the rest, he cited the world-weary author of Ecclesiastes: "All is vanity."

> Those careerists ambitious for glory by building vain buildings ought to look this way if they want to obtain the measure of eternal glory, and work at building churches in honor of divinity, put their effort into erecting hospices for receiving poor pupils; let them erect colleges to train youth in every virtue and fine art. And these are not Towers of Nimrod, but the creations of pious minds, whose cornices touch the very heavens, that is, which prepare a stairway for them to the apex of eternal joy, beyond which nothing greater remains to be desired. (P. 23)

> *I made me great works; I builded me houses; I planted me vineyards:*
> *I made me gardens and orchards, and I planted trees in them of all kind of fruits:*
> *I made me pools of water, to water therewith the wood that bringeth forth trees:*
> *I got me servants and maidens, and had servants born in my house; also I had great possessions of great and small cattle above all that were in Jerusalem before me:*
> *I gathered me also silver and gold, and the peculiar treasure of kings and of the provinces: I gat me men singers and women singers, and the delights of the sons of men, as musical instruments, and that of all sorts.*
> *So I was great, and increased more than all that were before me in Jerusalem: also my wisdom remained with me.*
> *And whatsoever mine eyes desired I kept not from them, I withheld not my heart from any joy; for my heart rejoiced in all my labour: and this was my portion of all my labour.*
> *Then I looked on all the works that my hands had wrought, and on the labour that I had laboured to do: and, behold, all was vanity and vexation of spirit, and there was no profit under the sun.*
> (Eccl. 2:4–11)

70. Galileo Galilei (1564–1642). *Discorso . . . intorno alle cose, che stanno in sù l'acqua, ò che in quella si muouono*. Florence: Appresso Cosimo Giunti, 1612. Berlin Collection

Kircher used Galileo's *Discourse on Floating Bodies* (*Discorso . . . intorno alle cose, che stanno in sù l'acqua*) to bolster his own arguments about the construction of Noah's Ark, first presented as a lecture during the centenary celebrations of the Society of Jesus in 1640:

> My immense fervor knew no restraint, so that it drove me so far as to take up the challenge of demonstrating without further delay to the public assembly of the Mathematical faculty the unseen Structure of the [Ark], as it had been established according to Geometric, Static, and Architectural principles. (Preface from Athanasius Kircher, *Arca Noe* [Amsterdam: Apud Joannem Janssonium à Waesberge, (1675)], P. ** 2 recto)

71. Photograph reproduced from Athanasius Kircher (1602–1680). *Turris Babel*. Amsterdam: Ex officina Janssonio-Waesbergiana, 1679.

According to Kircher, this diagram illustrates that an excessively high Tower of Babel knocks the earth from its proper place at the center of the universe. However, it is clear that the argument only makes sense if the earth is being knocked *out of orbit*: if the earth were the fixed center of the cosmos, it would not move under any circumstances.

In 1675 Kircher published *Noah's Ark* (*Arca Noe*) for Charles XII of Spain, the mentally retarded Hapsburg boy-king who was so inbred that his mother was also his first cousin (his father, Philip IV, had married his own niece). Somewhere between fairy tale, adventure book, and scientific treatise, *Noah's Ark* acted as a prelude to the *Tower of Babel* of four years later, not only because the two stories followed upon one another in Genesis, but also because they both addressed the issue of technology, whose advances Kircher sought to justify.

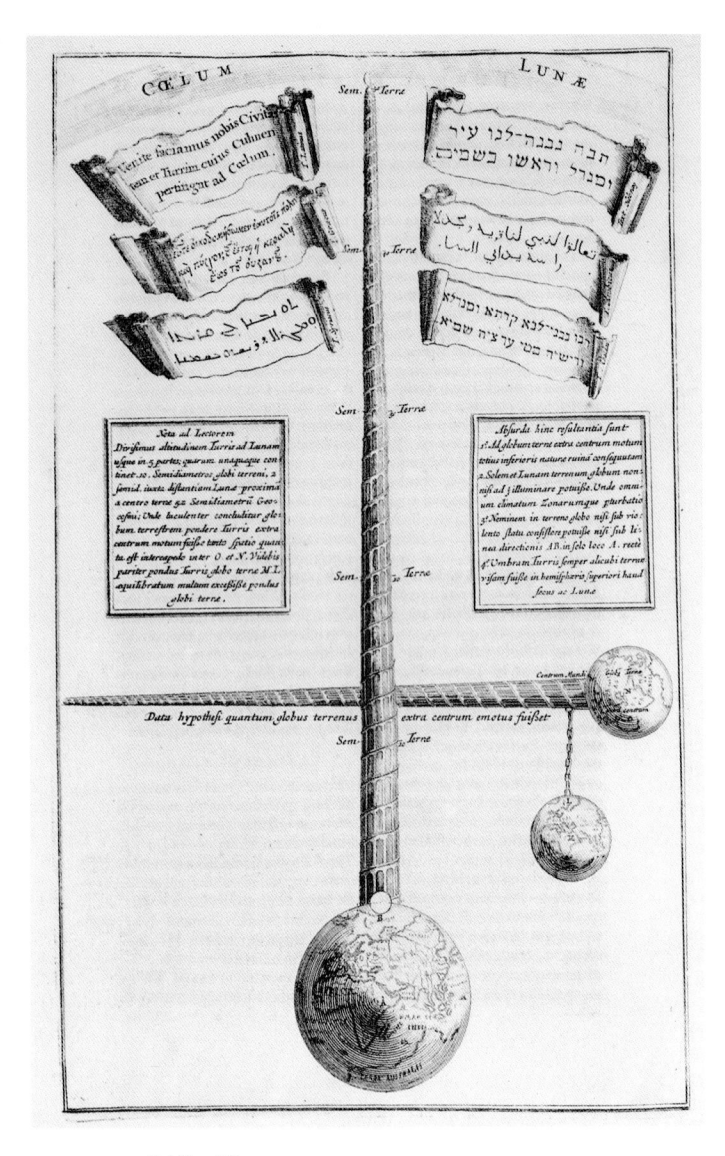

CAT. 71

72. Photograph reproduced from Athanasius Kircher (1602–1680). *Arca Noe*. Amsterdam: Apud Joannem Janssonium à Waesberge, [1675].
Courtesy of University of Chicago Department of Art History Slide Collection

The flood in full cataclysm must have appealed to Kircher's passion for nature's violent extremes, whereas the ark represented the calm serenity of God's order. An age-old symbol of the Church as a refuge from the world's turmoil, Noah's Ark had special appeal during the religious upheavals of the seventeenth century.

> Now if the Readers apply their minds a little more deeply to admiring the construction of this artifact, they will clearly observe that there was never its like in all of human history, nothing more worthy of human attention among all things, not only human but also Divine, since the restoration of the World, except that greatest event, Christ's advent in the flesh. Nor should this seem strange to anyone, for this Ark was dictated by God Himself as Architect to His servant Noah, according to each and every law of symmetry, laws prescribed, methods and principles of construction inspired as God taught them. . . . For this was a real and truthful compendium of the whole living World, a refuge for the whole Globe during the Flood, a new seedbed of humankind and animals after the world's destruction, in which each kind of man and animal was preserved and could be recovered, a new source for the world reborn. (P. 1)

# The Artificial Memory and the Combinatory Art

KIRCHER'S JESUIT EDUCATION emphasized strengthening the memory by means of the *Spiritual Exercises* of Ignatius Loyola, and by a technique passed down from ancient Greek and Roman times: the artificial memory. Students learned to "place" ideas as imaginary figures within elaborate architectural structures. The medieval Catalan mystic Ramon Llull (1232?–1316) adapted this system by replacing architecture with wheels, on which information could likewise be stored in imagery. Rotating the wheels created an immense number of combinations, and Llull called his technique the Combinatory Art. In the later sixteenth century, Giordano Bruno became an outstanding practitioner of Llull's combinatory method, performing astounding feats of memory as a way of advertising the efficacy of his version of the art. Kircher, in turn, attempted to apply the same system to the Jesuits' missionary vocation. Using a series of twenty-seven abstract symbols, he devised a universal language by which all the important aspects of Church teaching could be communicated no matter what language missionaries or native peoples might originally have spoken. In effect, his Great Art of Knowing was a first step toward symbolic logic.

73. Ramon Llull (1232?–1316). *Ars magna, generalis et vltima*. Frankfurt: Typis Ioan. Saurii, impensis Cornelii Sutorii, 1596.

The treatise shown here, Llull's *Great, General and Ultimate Art* (*Ars magna, generalis et ultima*), served as Kircher's model for his own *Great Art of Knowing*. Like Giordano Bruno and Kircher after him, Llull defended his Combinatory Art against suspicious critics by presenting it as a spiritual exercise:

> We created this Art in order to understand and love God, so that the human intellect can ascend artificially toward the knowledge of God and as a consequence into love, because the intellect can do with artifice what it cannot do without artifice, so long as divine grace and wisdom act as mediators. And because the more God is understood, the more He is loved, therefore this Art brings it about that God is loved as much as possible. (Ramon Llull, *Opera Latina 134. Ars compendiosa Dei . . .* , ed. by Manual Bauzà Ochogavia [Turnhout: Brepols, 1985], p. 17)

74. Ramon Llull (1232?–1316). *Opera*. Strasbourg: Sumptibus hæredum Lazari Zetzneri, 1651.

A comment by Athanasius Kircher in his 1669 book on memory systems, the *Great Art of Knowing* (*Ars magna sciendi*), shows that he had read precisely this 1651 edition of Llull's complete works; he stated that many scholars have commented on Llull, including Cornelius Agrippa of Nettesheim and "Jordanus," who is clearly Giordano Bruno. Both are included in this volume, safely published in Strasbourg, where the fact that both Agrippa and Bruno were on the *Index of Prohibited Books* had no effect.

> Just as a hand joined to an arm, a foot to a leg, or an eye to a head, are more recognizable than when they are separated, likewise, with parts and whole species, let nothing be set aside and out of order (which is utterly simple, perfect, and beyond number in the primal Mind) if we intend to connect them with each other and unite them: what then can we not understand, remember, and do? (Giordano Bruno, *De umbris idearum* [1582], Conceptus XV. P)

75. Athanasius Kircher (1602–1680). *Ars magna sciendi: in XII libros digesta*. Vol. 2. Amsterdam: Apud Joannem Janssonium à Waesberge & viduam Elizei Weyerstraet, 1669.
From the Library of Richard P. McKeon

The page displayed here from the *Great Art of Knowing* (*Ars magna sciendi*) shows a set of metaphysical syllogisms. In Kircher's scheme, whose terms are ultimately borrowed from Ramon Llull, B. stands for *bonitas*, goodness; M. for *magnitudo*, magnitude; D. for *duratio*, duration; P. for *potestas*, power; S. for *sapientia*, wisdom; Vo. for *voluntas*, will or love; Vi. for *virtus*, virtue; Ve. for *veritas*, truth; G. for *gloria*, glory. The pictorial symbols are nearly self-explanatory: the triangle represents God as the Holy Trinity, the cherub head represents angels, the circular diagram the cosmos, the square the four elements, the human figure humanity, followed by the categories of animal, vegetable, mineral, and number. Thus the first syllogism on the left side of the page states:

> Every Great Good (*Bonitas + Magnitudo*) shares [literally, "is diffusive"] of itself. Every divine, angelic, cosmic, elemental, human, animal, vegetable, mineral, and numerical entity is a Great Good.
>
> Therefore every divine, angelic, cosmic, elemental, human, animal, vegetable, mineral and numerical entity shares of itself. (P. 346)

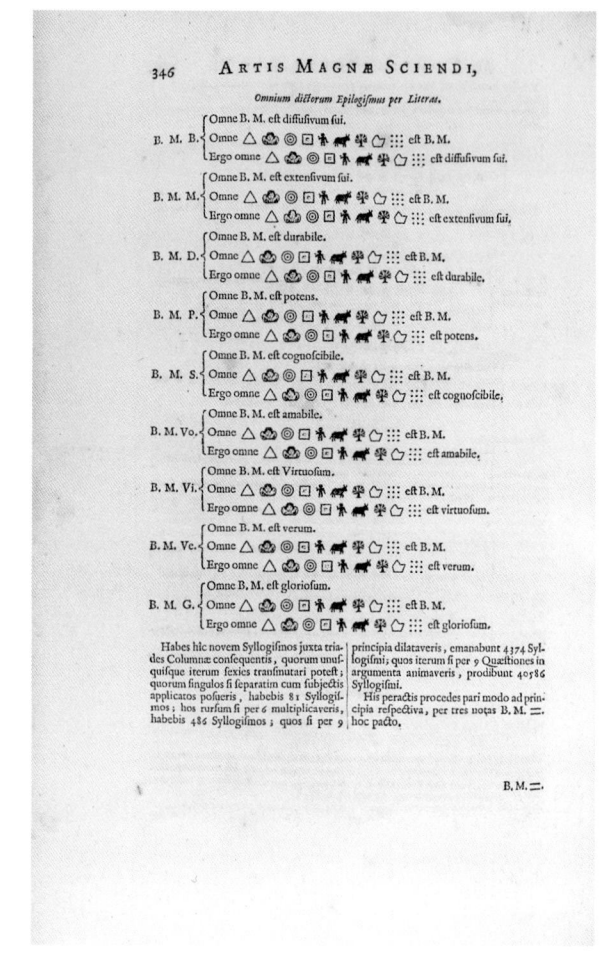

CAT. 75

The opposite page uses Kircher's symbols for relationship: the equal sign means "differs"; the heart means "agrees" (*concordat*—in Latin "is of like heart"); the black and white dots mean "are opposed"; alpha means "causes"; the circle with a center means "mediates"; omega means "has an end"; MA "is greater," from Latin *major*, "greater"; AE "is equal," from Latin *aequalis*; Min. "is lesser," from Latin *minor*, "lesser".

76. Photograph reproduced from Athanasius Kircher (1602–1680). *Ars magna sciendi: in XII libros digesta.* Vol. 2. Amsterdam: Apud Joannem Janssonium à Waesberge & viduam Elizei Weyerstraet, 1669. From the Library of Richard P. McKeon

The title page to the second volume of Kircher's *Great Art of Knowing* shows a classic memory palace, filled with images bearing the author's twenty-seven universal symbols. Unlike the abstract wheels of Llull and Bruno, this imaginary building has all the specific detail that Kircher so loved in the world itself.

CAT. 76

77. Quirin Kuhlmann (1651–1689). *Epistolæ duæ, prior de arte magnâ sciendi.* Leiden: Lotho de Haes, 1674. Berlin Collection

The Silesian poet and religious fanatic Quirin Kuhlmann was only twenty-three when he published his correspondence with the seventy-two-year-old Kircher, whose help he solicited for the publication of a promised masterwork on the Combinatory Art. Kircher's replies to the young firebrand are polite, guarded, and prophetic; he warned Kuhlmann about the dangers of "unconsidered writing" in an age of Inquisition, exhorting him to "use the necessary caution with secret matters before they see the light" (p. 43). Kuhlmann, however, did no such thing; as the incongruous appendix to a list of Kircher's publications, he exhorted the reader to heed the apocalyptic prophecies of an Amsterdam mystic, Jacob Rothe. Rothe was only one of several contemporary prophets to win Kuhlmann's endorsement. Initially a follower of the Rosicrucians, Kuhlmann would eventually establish his own sect, the Jesuelites, whose activities he moved to Moscow in 1686, where they aroused the immediate suspicions of Tsar Peter the Great. Tried for heresy (against the Orthodox Church), Kuhlmann was burned at the stake in Moscow in 1689. His masterwork on the Combinatory Art seems never to have progressed beyond the sketch he presented here in a few bombastic pages.

78. Gaspar Schott (1608–1666). *Pantometrum Kircherianum, hoc est, instrumentum geometricum novum.* Würzburg: Excudebat Jobus Hertz, 1660. Berlin Collection

Gaspar Schott described Kircher's universal surveying tool, the *Pantometrum*, as follows:

> More than once in Rome, with the greatest pleasure to my spirit, I saw the Inventor instruct noble youths, Barons, Counts, Dukes, Princes, some native, some foreign, who had hardly knocked at the door of the Mathematical Sciences, who lacked the basic elements of Arithmetic; these, nonetheless, [he taught] to use the Instrument as skillfully, and to solve the Geometric and Geodetic problems set for them as expertly, as he could expect a professional surveyor to do. (p. ).( ).( ).(3r)

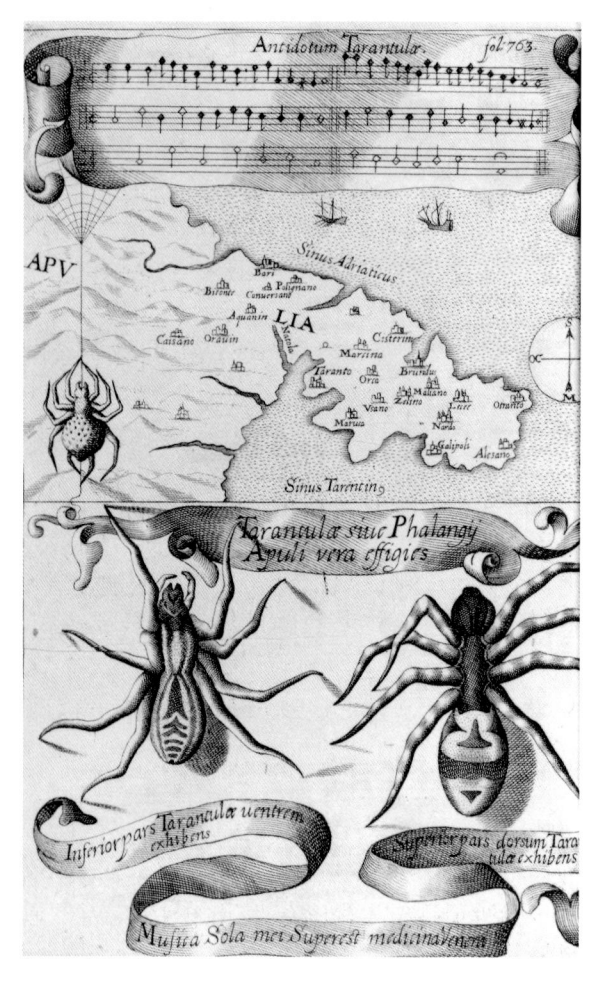

CAT. 79

79. Athanasius Kircher (1602–1680). *Magnes, siue De arte magnetica*. Cologne: Apud Iodocum Kalcouen, 1643.
John Crerar Collection of Rare Books in the History of Science and Medicine

The German reprint of Kircher's *The Magnet, or the Magnetic Art* was published only two years after the first Roman edition, an indication of the book's immediate international popularity. Here Kircher used the universal powers of magnetism to explain how dancing the tarantella can draw out the toxins of a tarantula bite. Typically, his illustration supplies a vivid image of a tarantula, a musical score, and a nostalgic view of a volcanic landscape in Southern Italy, a fond reminiscence of his trip there in 1637–38.

# *The Universal Language I: Jesuit Missionaries*

As an order deeply concerned with education and with missionary work, the Jesuits looked for efficient ways to convey the Christian message to a wide variety of people, hoping to find some kind of universal language comprehensible to all. Early experiments with communication by gesture produced dramatic results among the deaf; in fact, American Sign Language descends directly from these early Jesuit researches. Athanasius Kircher's investigations of Egyptian hieroglyphs were motivated in part by his hope that he could invent a universal picture-writing. Taking a different but equally effective approach, missionaries to China, like Matteo Ricci and Adam Schall von Bell, distinguished themselves by their ability to adapt to their surroundings; through their mastery of local language and customs, both rose to the top of the Chinese imperial court, Schall von Bell even handily surviving the change from Ming to Qing dynasties.

80. Pietro della Valle (1586–1652). *Viaggi di Pietro della Valle il Pellegrino*. Vol. 2. Venice: Appresso Gio. Battista Tramontino, 1681.

The Roman aristocrat Pietro della Valle traveled in Turkey, Persia, and India in the second decade of the seventeenth century, sending back effusive letters along the way with vivid descriptions of the lands and peoples he saw. Published repeatedly throughout the century, these letters were immensely popular with diverse readers, including Athanasius Kircher. In Cairo, della Valle obtained the medieval Arabic manuscript that would eventually pass into Kircher's hands and form the basis for three of his books: the *Coptic Forerunner* (*Prodromus Coptus*), the *Egyptian Language Restored* (*Lingua Aegyptiaca restituta*), and the *Egyptian Oedipus* (*Oedipus Aegyptiacus*). Della Valle also had his own way of assimilating to the local culture, one not open to Jesuit missionaries: he married a Persian woman, whom he called "Signora Maani." Writing from Isfahan in 1621, he reported on some of the luxuries that life in Persia had brought him, such as ice cream and Persian cats:

It not only refreshes at once, like snow, and better, but also makes you cheerful with the pretty sight of its transparency; it can't be expressed how much it delights. In Isfahan there are many, many of these ice factories, because the city consumes a tremendous quantity every year, and I wanted to describe the story in detail and at length, because it seems worth imitating in our countries, for which reason I want it to be well known in Italy. (p. 164)

I love my country, and want to enrich it with whatever I find in other countries that is good or beautiful. So . . . here I have seen a beautiful breed of Cats, next to which our beloved Soriani are nothing, and finally I want to bring this breed to Rome. In size and form they are ordinary Cats; their beauty lies in their color and their fur. The fur is very fine, lustrous and soft, like silk. . . . The most beautiful thing they have is their tail, which is very large and full of fur, so long that it looks like a squirrel's [tail], and just like squirrels they arch them over their backs like plumes, which produces a beautiful effect. They are also very tame, so much so that Signora Maani [della Valle's wife] can't resist letting them in bed, even under the sheets [della Valle's cats were named Ambar, Caplan, Farfanicchio, and Ninfa]. (pp. 164–65)

CAT. 82

81. Matteo Ricci (1552–1610). *De Christiana expeditione.* Augsburg: apud Christoph. Mangium, 1615.
Ernst Wilhelm Hengstenberg Collection

Matteo Ricci's *Christian Expedition to China* (*De Christiana expeditione*) gave a gripping account for European readers of his experiences in China. The book is open to the page where Ricci begins to discuss how he assumed Chinese dress and gained entree to learned circles by teaching the art of memory. Ricci also wrote extensively in Chinese for the people among whom he would spend the rest of his life.

82. Athanasius Kircher (1602–1680). *China monumentis qua sacris quà profanis . . . illustrata.* Amsterdam: Apud Joannem Janssonium à Waesberge & Elizeum Weyerstraet, 1667.

This image from Kircher's *China Illustrated in Monuments* (*China monumentis . . . illustrata*) shows Matteo Ricci in Chinese dress with his Chinese associate, the mandarin convert Paul Li.

83. Photograph reproduced from Athanasius Kircher (1602–1680). *Turris Babel.* Amersterdam: Ex officina Janssonio-Waesbergiana, 1679.

Confined as he was to Italy, Athanasius Kircher dreamed of foreign places and was an avid reader of Pietro della Valle's travel letters. In his treatise on the Tower of Babel, he made extensive use of della Valle's reports of explorations among the ruins of Babylon; like most modern archaeologists, Kircher assumed that the biblical tale of the Tower of Babel referred to the Babylonian ziggurat. This engraving of its ruins ranks among the finest illustrations in his works. For Kircher the confusion of languages and scattering of peoples that followed on the tower's destruction ranked as one of the defining moments of human history; his search for a universal language was an attempt to heal that primordial injury.

CAT. 83

# The Universal Language II: Egyptian

HOWEVER FUTILE THEY LOOK today, Athanasius Kircher's researches into ancient Egyptian represented a great advance over those of his chief predecessor, the Renaissance scholar Giovanni Pierio Valeriano (1477–1560). Valeriano's interpretations of hieroglyphs rely largely on a late antique (fourth century C.E.) treatise attributed to a Greek-speaking Egyptian named Horapollo; the treatise's discovery in the fifteenth century had been the cause for great excitement. For Kircher, however, the insights provided by Horapollo were insufficient; he wanted to know how the Egyptian language worked grammatically. He believed that Coptic, the liturgical language of Egyptian Christians, must hold the key, and he was right. Hoping that wider knowledge of Coptic would lead to a breakthrough in the study of Egyptian, Kircher translated into Latin and published a medieval Coptic-Arabic grammar book; he had acquired the original manuscript from the traveler Pietro della Valle. Again, Kircher's instincts were right, but the final decipherment of hieroglyphs would also require the help of the Rosetta Stone, whose discovery lay more than a century in the future. Appropriately, when Jean-Louis Champollion finally began to read the Rosetta Stone's Egyptian inscriptions, his annotated copy of Kircher's *Coptic Forerunner* (*Prodromus Coptus*) (now in the Bibliothèque Nationale in Paris) was at his side.

84. Pierio Valeriano (1477–1560). *Hieroglyphica, seu de sacris Aegyptiorum.* Lyon: Apud Bartholomæum Honoratum, 1586.
Helen and Ruth Regenstein Collection of Rare Books

This pioneering work on the interpretation of hieroglyphs was immensely popular. Athanasius Kircher, an attentive reader, borrowed some of the book's clever marketing techniques: the generous size, the copious illustrations, the dedication of individual chapters to different patrons. Today these dedicatory prefaces to Valeriano's chapters constitute the book's most rewarding reading, offering vivid characterizations of a wide variety of important sixteenth-century intellectuals from many walks of life.

85. Athanasius Kircher (1602–1680). *Prodromus Coptus siue Ægyptiacus.* Rome: Typis S. Cong., 1636.
Bequest of Lessing Rosenthal

The *Coptic, or Egyptian, Forerunner, in Which Both the Origin, Age, Vicissitude, and Inflection of the Coptic or Egyptian, Once Pharaonic, Language, and the Restoration of Hieroglyphic Literature Are Exhibited* (*Prodromus Coptus siue Aegyptiacus . . .*) represents Athanasius Kircher's first Egyptological work, published in Rome in 1636. It is dedicated to his first sponsor in the city, Cardinal Francesco Barberini, and was printed at the press of the missionary body known as the Sacred Congregation for the Propagation of the Faith, for Kircher believed that the ability to read hieroglyphs would further the missionary search for a universal language. In order to gain credibility, Kircher collected congratualatory poems in twelve languages, including this offering in biblical Hebrew:

> Hebrew Song
>
> Every tongue in every province
> From east to west
> Both Hebrew and Roman; they are not enough for you.
> The Coptic tongue is your tongue
> Borne up into the light out of darkness
> In twelve languages it is exalted.
>
> (P. +++2 verso)

The censor's statement, by his fellow Jesuit Melchior Inchofer, was unusually effusive:

> I censored the book of Father Athanasius Kircher, S.J., which is known as *Prodromus Coptus, seu Aegyptiacus*, according to religious usage and order. In it, I not only found nothing to offend the Catholic Religion and repugnant to good morals, but on the contrary, many arguments brought forth ingeniously from the hidden sanctuaries of holy antiquity and the mysteries of the Egyptians, which, together with genuine knowledge of many languages as well as erudition in secret exotic matters, drunk up from the same sources, which supply strength and confirmation to orthodox truth, toppling heresies and errors. A worthy beginning from which we may anticipate what will follow.
>
> (P. ++2 verso)

86. *Table of Isis, or Mensa Isiaca*. Engraving from Athanasius Kircher (1602–1680). *Oedipi Aegyptiaci tomus secundi pars alteri*. Rome: Ex typographia Vitalis Mascardi, 1653. Vol. 2 of *Oedipus Aegyptiacus*. Rome: Ex typographia Vitalis Mascardi, 1652–54.

The artifact shown in this engraving from Kircher's *Egyptian Oedipus* is a bronze tabletop inlaid with silver figures, now one of the treasures of the Egyptian Museum in Turin. It was discovered in Rome in about 1525, on the site of the ancient Roman Temple of Isis, and purchased by the Venetian writer and future cardinal Pietro Bembo, who gave it the name "Tabula Bembina" ("Bembo's Table"). Kircher called it the "Mensa Isiaca" ("Table of Isis"). From the moment of its discovery, its figures and hieroglyphic inscriptions became chief subjects of study by would-be Egyptologists, including Kircher, who carried a sketch of the Mensa Isiaca with him everywhere. Unfortunately, we now know that this bronze offering table was made by a Roman craftsman for a Roman customer; the hieroglyphs

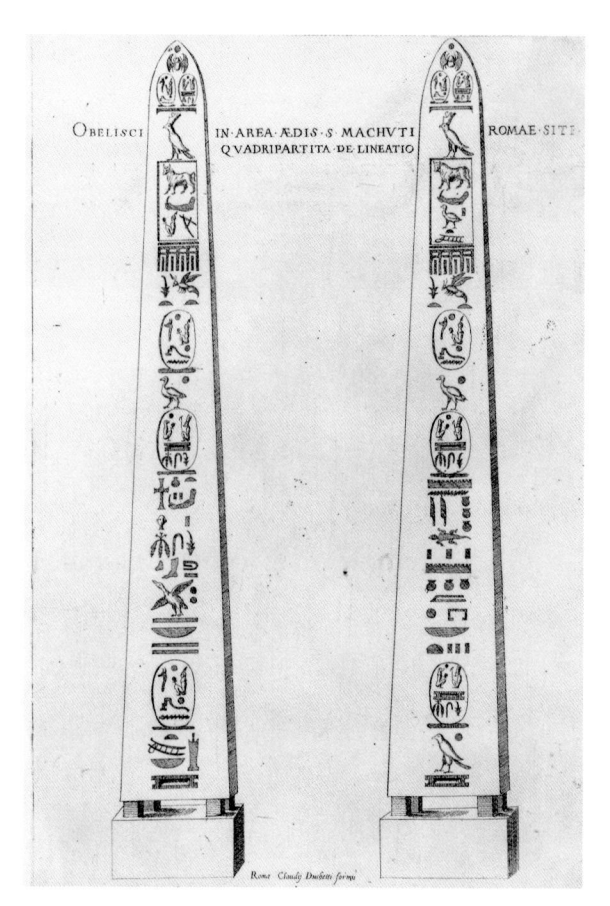

CAT. 87

are nonsense, a Roman's idea of what hieroglyphs should look like. The hours that people like Bembo, Pierio Valeriano, and Kircher spent in studying the figures could teach them nothing about the ancient Egyptian language. Fortunately, they also took great aesthetic delight in the table's wonderful cosmpolitan amalgam of Roman and Egyptian taste.

87. [Etienne Duperac (ca. 1525–1604)], engraver. *Obelisci in area ædis S. Machuti Delineatio*. Rome: Claudij Duchetti, [1581–86]. Engraving added to *Speculum Romanae magnificentiae*. Rome: Antonio Lafreri, ca. 1544–77.

The obelisk of San Macuto (Saint Maclovius) was discovered on the grounds of the former Temple of Isis, probably at the end of the fourteenth century. It is now displayed in front of the Pantheon, where it was installed in 1711 by Pope Clement XI.

# Kircher's Obelisks: The Pamphili Obelisk and Bernini's Fountain of the Four Rivers, Piazza Navona

ATHANASIUS KIRCHER MAY have played a crucial role in the creation of Gianlorenzo Bernini's masterpiece, the Fountain of the Four Rivers in Piazza Navona (completed in 1648). He supplied concise Latin translations of the inscriptions on the four sides of the fountain's central obelisk; they can be seen engraved on plaques beneath the obelisk itself. Moreover, the very form of the fountain reflects Kircher's concept of geology; he believed that mountain ranges stood hollow over vast underground reservoirs in which all the rivers of the world had their origins. Bernini's design of a hollow mountain is therefore not only sensible from a structural point of view, but also an expression of the latest in scientific thinking.

88. Athanasius Kircher (1602–1680). *Obeliscus Pamphilius*. Rome: Typis Ludouici Grignani, 1650. Given by Sinai Congregation

Published to honor Pope Innocent X in the Jubilee Year of 1650, this treatise on the obelisk in Rome's Piazza Navona also contains the clue to reading all of Kircher's work on hieroglyphs: profound religious truths must be couched in symbolic language so that they do not enter the ears of the common crowd.

CAT. 88

CAT. 91

89. [Giovanni Antonio Dosio (1533–1609?)], artist and engraver. *Extra portam Capenam . . . circus.* [Rome or Florence, 1569]. Engraving added to *Speculum Romanae magnificentiae*. Rome: Antonio Lafreri, ca. 1544–77.

In the early fourth century C.E., the emperor Maxentius (reigned 306–312), a rival of Constantine, transferred an obelisk erected in the Roman precinct of Isis by the emperor Domitian (reigned 82–96) to his own circus outside the walls of Rome, built in honor of his deceased son Romulus. There it lay in pieces until Pope Innocent X (reigned 1644–1655) decided to make it the centerpiece of a fountain in front of his family palazzo in Piazza Navona, on the site of the former Circus of Domitian. The transfer was completed in 1648.

90. Leone Allacci (1586–1669). *Summikta*. Cologne: Apud Iodocum Kalcovium, 1653.

The Greek-born scholar (and future Vatican librarian) Leone Allacci shared his friend Kircher's love of old manuscripts and linguistic studies. The essay on Coptic religious rites shown in this *Miscellany* (*Summikta*) of Allacci's works is inspired by and dedicated to Kircher.

91. Photograph reproduced from Athanasius Kircher (1602–1680). *Mundus subterraneus, in XII libros digestus.* Vol. 1. Amsterdam: Apud Joannem Janssonium & Elizeum Weyerstraten, 1665. John Crerar Collection of Rare Books in the History of Science and Medicine

Athanasius Kircher believed that the world's mountain ranges concealed vast water reservoirs, which he called *hydrophylacia* ("water reservoir" in Greek), one of which is shown here. Bernini's Fountain of the Four Rivers seems to exhibit the same geology.

# *Pyramids and Mummies:* Kircher's Initiatory Sphinx

KIRCHER'S LATE WORK *The Initiatory Sphinx* (*Sphinx mystagoga*) (1676) shows how international his reading public had become by the end of his life; it was written for a learned French collector who had brought two mummy cases from Egypt and hoped for a translation of their hieroglyphic inscriptions. The book was printed simultanously in Rome by the house of Vitale Mascardi and in Amsterdam by Jansson and Weyerstraet. For some readers, evidently, the interpretation of hieroglyphs in Kircher's *Egyptian Oedipus* remained convincing two decades after its publication.

92. Athanasius Kircher (1602–1680). *Sphinx mystagoga.* Amsterdam: Ex officina Janssonio-Waesbergiana, 1676.

The mummies Kircher interpreted in *The Initiatory Sphinx* (*Sphinx mystagoga*) had been found in Memphis, near modern Gizeh. Neither Kircher nor his engraver had ever seen real Egyptian pyramids, with their relatively squat proportions (the width of the bases at Gizeh are equal to the pyramids' height). They had only seen the tall, slender Roman version (twice as tall as it is broad) erected by Gaius Cestius in 12 B.C.E. and restored by Pope Alexander VII in 1663. This imaginative view of Memphis thus looks decidedly strange to modern eyes; furthermore, like many of his contemporaries, the engraver did not (or could not) make a clear distinction between pyramids and obelisks.

93. Photograph reproduced from Athanasius Kircher (1602–1680). *Sphinx mystagoga.* Amsterdam: Ex officina Janssonio-Waesbergiana, 1676.

Kircher and his engraver were evidently fascinated by the various methods used to wrap Egyptian mummies in linen bands, and by the intricate beauty of the patterns they formed.

94. Athanasius Kircher (1602–1680). *Oedipi Aegyptiaci, tomus III.* Rome: Ex typographia Vitalis Mascardi, 1654. Vol. 3 of *Oedipus Aegyptiacus*. Rome: Ex typographia Vitalis Mascardi, 1652–54.

A real sensitivity to the difference between the artistic conventions of seventeenth-century Europe and those of other cultures is evident here and in some of the engravings for Kircher's *China Illustrated in Monuments*; this illustration shows a genuine effort to convey both the flatness of ancient Egyptian style and the expressively elongated arms of a protective deity.

95. [François Perrier (1590?–1656?)], artist and engraver. *Sepulchrum C. Cestii.* Rome: Claudii Ducheti, 1582. Engraving added to *Speculum Romanae magnificentiae.* Rome: Antonio Lafreri, ca. 1544–77.

In 12 B.C.E., a Roman magistrate named Gaius Cestius decided to have his tomb along the Via Ostiense fashioned in the form of a pyramid. It was the heyday of Emperor Augustus, who had defeated the fleet of Cleopatra and Marc Antony nineteen years before, and Egypt was now annexed to Rome. Unlike Egyptian pyramids, Cestius' creation is tall and slender, twice as tall as it is wide at the base, and it is made of marble veneer over a concrete core. Kircher's friend and patron Pope Alexander VII had restored it only recently when this image was produced, certainly with Kircher's Egyptological interests in mind.

CAT. 93

# *Kircher's* Egyptian Oedipus

ATHANASIUS KIRCHER'S massive four-volume *Egyptian Oedipus* (*Oedipus Aegyptiacus*) represented the culmination of his Egyptological studies. It took over three years to print, with its hundreds of illustrations—although the title pages give the dates 1652–54, parts of the second volume were not finished until 1655. Kircher knew that his reputation as a translator of Egyptian hieroglyphs would stand or fall on this work, and that is why its first seventy pages consist of collected compliments from scholars all over the world, writing in every possible language, from Chinese to Coptic, Ethiopian to English, German to Greek. Kircher drew most of his illustrations from the Egyptian artifacts visible in Rome, on a large scale like the Pyramid of Gaius Cestius, or on a small scale like the two objects depicted here in contemporary engravings: a relief of a procession in honor of Isis, and an Egyptian canopic jar (designed to hold the entrails of mummified bodies), on exhibit in the Capitoline Museum since its establishment by Pope Sixtus IV in the fifteenth century—the very first public museum.

96. Athanasius Kircher (1602–1680). *Oedipus Aegyptiacus.* Vol. 1. Rome: Ex typographia Vitalis Mascardi, 1652.

Athanasius Kircher regarded his *Egyptian Oedipus* as a major achievement. Its title page shows him figuratively as a young Greek hero unlocking the riddle of the Sphinx, but the decades of work that went into the book better fit a middle-aged professor, as Kircher was by 1652.

> So here, benevolent Reader, you have a Work born of twenty years of a continuous mental firestorm. (P. c verso)

Although he credited ancient Egypt with great sophistication, Kircher was horrified at the Egyptians' propensity to worship animals:

> I am truly amazed that it was possible for people otherwise of sound mind ever to have accepted, let alone approved, such insane and fanatical hallucinations. (P. 359)

Among the seventy pages of laudatory correspondence that precede the text of the *Egyptian Oedipus*, Kircher included a dedicatory poem by the English expatriate in Rome, James Alban Gibbes, M.D., and Poet Laureate (1611–1677):

> Rome shows in whole and parcels all the Rubble
>    Of wasted Aegypt, giving pleasant trouble
> And most sweet rack to witts, to know, and see
>    The mangled parent of Antiquitie:
> Aegypt, mother of arts, where better might
>    Then here, i' th' lapp of science, take delight

> Gather'd in Rome, dismembered? perhaps too
>   Appeare far brighter, then did ever doe . . .
> This is the United sense of th' Universe
>   Though differing tongues it many ways reherse.
>                                    (P. +++++ 2 recto)

97. [Pietro Santi Bartoli (ca. 1635–1700), in part after François Perrier (1590?–1656?), engraver]. *Isidis Pompa*. [Rome: Domenico de Rossi, 1693]. Engraving added to *Speculum Romanae magnificentiae*. Rome: Antonio Lafreri, ca. 1544–77.

The Egyptian deity Isis, with her benevolent roles as loving mother and loyal wife, grew increasingly popular as contacts between Egypt and Rome expanded in the first century B.C.E. Although the Roman Senate periodically moved to suppress the cult of Isis, it retained its importance until the Christian era, with its most important sanctuary situated in the Campus Martius, next to the present-day Piazza Venezia—and the Jesuits' Collegio Romano. Many of the roles that Isis played in ancient Romans' spiritual lives were assumed by the Virgin Mary. This Roman relief, executed in Roman style, shows Roman devotees of the goddess shaking her distinctive bronze rattle, the sistrum. It may well have been found on the site of the ancient Isis precinct.

98. [Etienne Duperac (ca. 1525–1604)], engraver. [*View of the Canopic Vase*]. Rome: Claudij Duchetti, [1581–86]. Engraving added to *Speculum Romanae magnificentiae*. Rome: Antonio Lafreri, ca. 1544–77.

These basalt jars, designed to hold the entrails of Egyptian dead and executed by Egyptian artists in Egyptian style, were probably found on the site of the sanctuary of Isis in the Campus Martius. They had entered the collections of the Capitoline Museum by the early sixteenth century, and provided European artists with a firsthand example of ancient Egyptian sculptural style, a style adapted to the extreme hardness of a stone that had to be worked by slow grinding much more than chiseling.

99. Lucius Cornelius Europaeus [Melchior Inchofer; ca. 1585–1648]. *Monarchia solipsorum*. Venice: n.p., 1651.
Helen and Ruth Regenstein Collection of Rare Books

Not every scholar who followed Kircher's Egyptological studies was pleased with the results. Melchior Inchofer, who had been so complimentary when he censored Kircher's *Coptic Forerunner*, had lost his enthusiasm by 1645, when he penned this acid satire of the Jesuit Order, *The Monarchy of the Solipsists*. Caricatures of Kircher and his work seem unmistakable in the following passages from that diatribe:

> Then an Egyptian wanderer came in; sitting in the middle of the piazza on a wooden crocodile, he broadcast trifles about the Moon. He said that the Moon was nothing other than a huge wheel of cheese, reduced day by day at the banquets of the Gods until it was entirely consumed, and then, every month a new one, of the same size and shape, was put in its place. (P. 104)

> Philosophical works among them are more or less of this sort: "Does the scarab roll dung into a ball paradigmatically?" [see *Oedipus Aegyptiacus*, vol. II.2, P. 212]. "If a mouse urinates in the sea, is there a risk of shipwreck?" "Are mathematical points receptacles for spirits?" "Is a belch an exhalation of the soul?" "Does the barking of a dog make the moon spotted?" and many other arguments of this kind, which are stated and discussed with equal contentiousness. Their Theological works are: "Whether navigation can be established in imaginary space." "Whether the intelligence known as *Burach* has the power to digest iron." "Whether the souls of the Gods have color." "Whether the excretions of Demons are protective to humans in the eighth degree." "Whether drums covered with the hide of an ass delight the intellect." (P. 29)

100. Curzio Inghirami (1614–1655). *Ethruscarum antiquitatum fragmenta*. Frankfurt: n.p., 1637 [i.e. Florence: Amadore Massi, 1636].
Helen and Ruth Regenstein Collection of Rare Books

Melchior Inchofer, who wrote *The Monarchy of the Solipsists* as "Lucius Cornelius Europaeus," also wrote a pseudonymous denunciation of the young Tuscan nobleman Curzio Inghirami's forged Etruscan time capsules, called *scarith*, discovered in Volterra in 1634. Prodded by his powerful relatives and by the grand duke of Tuscany, Ferdinand II, young Inghirami published his *Fragments of Etruscan Antiquities* (*Ethruscarum antiquitatum fragmenta*) for a general reading public in a handsome folio volume with a false Frankfurt imprint to give it cosmopolitan flair. The "Volterran Antiquities" quickly became an arena in which scholars all over Europe began to parade their expertise as antiquarians. At the time, Athanasius Kircher, who might have been expected to comment, was expecting an expensive Arabic type font from Grand Duke Ferdinand, and therefore held his tongue. Only in 1650, with his Arabic type firmly in hand, did he dismiss Inghirami's forgeries in a single sentence of his *Pamphili Obelisk*, the book in which his new Arabic font was used for the first time.

101. Photograph reproduced from Athanasius Kircher (1602–1680). *Obeliscus Pamphilius*. Rome: Typis Ludouici Grignani, 1650.
Given by Sinai Congregation

Melchoir Inchofer may have had this image of a dung beetle rolling a cosmic orb in mind when he wrote, "Does the scarab roll dung into a ball paradigmatically?" in his satire, *The Monarchy of the Solipsists*.

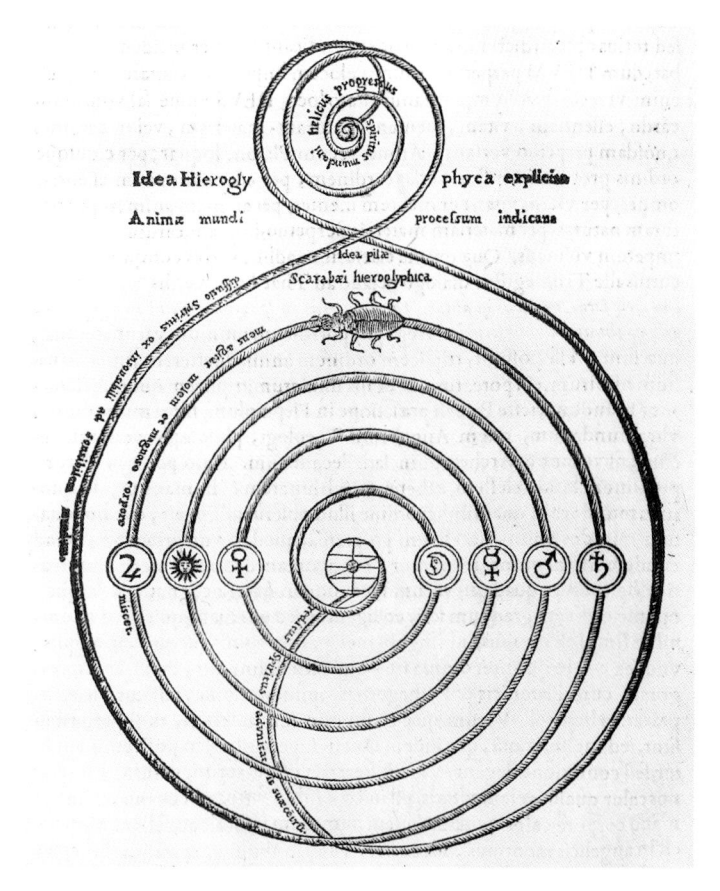

CAT. 101

# *Latium and* Latium, *1671*

AFTER HIS RETURN FROM SOUTHERN ITALY, SICILY, AND MALTA in 1638, Athanasius Kircher stayed in the environs of Rome for the rest of his life. However, his restless energy took him all over the countryside of Latium, the region around Rome, to visit ancient sites like the ruins of Emperor Hadrian's Villa; natural wonders like the waterfalls of Tivoli and the volcanic lakes of Albano and Nemi; and sacred retreats like the monastery of Grottaferrata and the little medieval shrine to Saint Eustace at Mentorella, to which he showed special devotion. The ruined grandeur of ancient Roman aqueducts and villas, together with the visible evidence of geologic change in the landscape, prompted Kircher to meditate on what he called the "vicissitude" of the world, its eternal propensity to change:

> All these things I frequently examine with the most intimate scrutiny of my mind, and soon, with good reason, that praiseworthy pronouncement of Ecclesiastes comes to mind: *Vanity of vanities; all is vanity* [Eccl. 2:2]. . . . There is nothing, unless it be eternal, that can be stable and lasting, which shall be clearly shown in the course of this work as if in a bright mirror. Indeed, if you see, horrified, what are but the lonely corpses, scattered everywhere, of once flourishing and powerful cities, you will marvel at what today are no more than chaotic heaps of stones, from villas and palaces built with supreme magnificence for every practice of pleasure, and where once there were the delightful retreats of Emperors, today you will observe the dens of beasts, serpents, owls, overgrown by thorns and thistles, and with wonderment you will discern that structures erected to last an eternity have in the course of a few centuries fallen into endless ruin; thus in this frailty of corrupt life nothing comes our way that is so splendid, magnificent, powerful, and strong that it should not be regarded as momentary in its destiny to change. (Athanasius Kircher, *Latium* [1671], preface)

CAT. 103

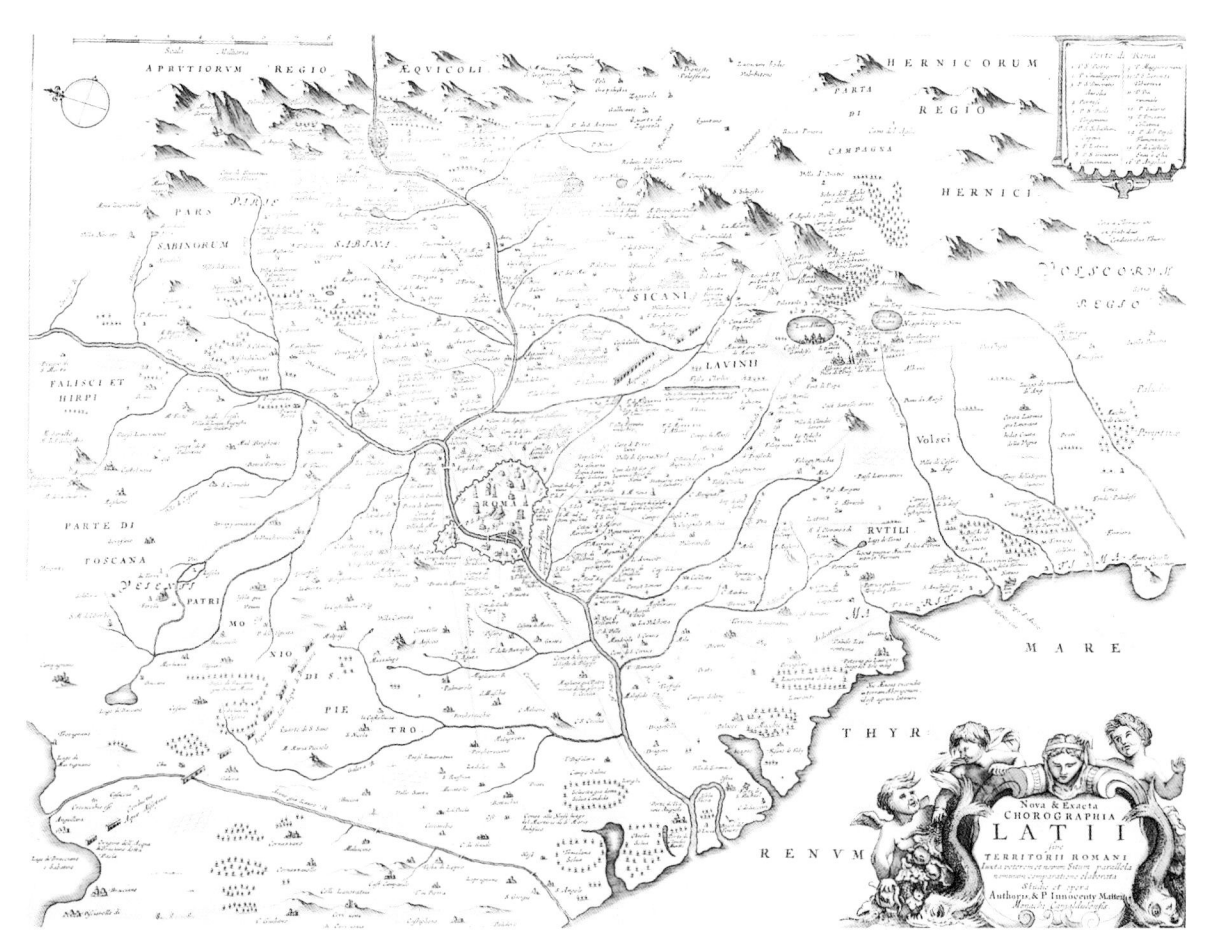

102. Pirro Ligorio (ca. 1510–1583). *Pianta della villa tiburtina di Adriano Cesare*. Rome: Stamperia di Apollo, 1751.
Berlin Collection

This plan of the villa built near Tivoli by the Roman emperor Hadrian (reigned 119–135) was first drafted by the Neapolitan architect and antiquarian Pirro Ligorio and reprinted for centuries thereafter; this version is from 1751. Ligorio's plan guided Kircher's own explorations of Hadrian's Villa.

103. Photograph reproduced from Athanasius Kircher (1602–1680). *Latium*. Amsterdam: Apud Joannem Janssonium à Waesberge & hæredes Elizei Weyerstraet, 1671.
Purchased from the Bequest of Louis H. Silver

Kircher's skill as a cartographer was denounced by two antiquarian readers of his *Latium*, who accused him (in a review of the book now preserved in the Vatican Library) of getting the distances between places entirely wrong. From a modern point of view, the map's interest lies in all the places that have since become engulfed by Rome's twentieth-century urban sprawl.

> At the top of the frontispiece . . . the arms of the City of Rome, SPQR, are reversed, with the bar going in the bastard direction, an unhappy prediction that the information about Latium will be bastardized rather than reported legitimately and infallibly, as the author magnificently promises the Pope in the dedication. (Vatican Library, MS Chigi N.III.82, 154r; written 1672 by R. Fab. and Marco Polo, "Antiquarii extra muros")

CAT. 103

104. Athanasius Kircher (1602–1680). *Historia Eustachio-Mariana*. Rome: Ex typographia Varesij, 1665.

In Mentorella, Kircher helped to restore a little medieval shrine marking the place where the knight Saint Eustace saw a vision of the cross between the antlers of a stag. Although he put implicit faith in prophetic visions—he had many himself—Kircher was reluctant to believe in miracles. Nature's own laws were wondrous enough.

> I say, those things happened by no principle or divine power or work of Benevolent Angels, because if all these things were ordered to propagate idolatry and superstitions in the minds of the Gentiles, then if they happened by Divine or angelic power, God and the Angels would be seen as cooperating by another means in the evil and the detestable sins that are committed in sacrifices. (Athanasius Kircher, *Latium* [1671], p. 40)

> [Portents are] like hieroglyphic symbols, swathed in enigmatic and allegorical meanings, which the Divine Wisdom records in Heaven, Earth, and the elements as if in a book and sets it before mortals to read; when they withdraw from the paths of Divine Will they are terrified by the threats held out before them, and turn back toward better fruits. (Athanasius Kircher, *De prodigiosis crucibus* [1661], p. 84)

105. Athanasius Kircher (1602–1680). *Phonurgia nova sive conjugium mechanico-physicum artis & naturae paranympha phonosophia concinnatum*. Kempten: Rudolphum Dreherr, 1673.

Kircher's *New Soundcraft* (*Phonurgia nova*) of 1673 describes his experiments with noisemakers in the countryside around Mentorella. This plate shows how he tested the volume of his horns and megaphones by blasting them in various directions, asking neighbors to wave handkerchiefs if they heard the noise, or to light candles if he made the experiments by night. Some of these instruments were as tall as the musicians hired to play them, and in the silence of the seventeenth-century rural landscape they really could be heard for miles.

106. Athanasius Kircher (1602–1680). *Physiologia Kircheriana experimentalis*. Amsterdam: Ex officina Janssonio-Waesbergiana, 1680.

Johann Stephan Kestler made many of Athanasius Kircher's machines at the Collegio Romano. This anthology of excerpts from Kircher's published works presents some of the most famous devices the two built over the years, including, as seen here, some of the kites, balloons, and noisemakers first described in a large work of 1650, *Universal Music-Making* (*Musurgia universalis*). The hot-air balloon in the woodcut illustration shows one of the pranks that Kircher played on the residents of the area around Mentorella, and from the frequency with which he repeated the image in his books, it must have been one of his favorites. The paper dragon was kept aloft by inflated bladders under which he placed lighted candles; when the candles burned down, the balloon went up in flames. Kircher's original sketch for this flying dragon, preserved in the National Library (Biblioteca Nazionale Centrale) in Rome, shows that its belly should read IRA DEI FUGITE—"Flee the wrath of God." He hoped that his practical jokes would rid people of superstition by showing that these unidentified flying objects operated according to nature's standard laws.

## *Censorship and the* Index of Prohibited Books

FROM THE OUTSET, ATHANASIUS KIRCHER'S INTELLECTUAL LIFE in Rome was hedged in by the curricular strictures of his own order, especially the Jesuit *Course of Study* (*Ratio studiorum*), adopted in 1599. He was equally constrained by the rule of the Inquisition, an institution founded in the late fifteenth century but grown much more active as Protestant creeds gathered favor in the powerful states of Northern Europe. Because so many of the Jesuit fathers acted as censors and examiners for the Roman Inquisition, the library at Kircher's own Collegio Romano contained a vast trove of forbidden books, but they were read there by well-trained defenders of the faith.

The *Index of Prohibited Books* was not always a carefully written or edited document. Each Inquisition issued its own list and had its own special concerns. For years, the Madrid and Lisbon editions of the *Index* perpetuated the same typographical error in crediting the works of Giordano Bruno of Nola, "Jordanus Brunus Nolanus," to "Jordanus Bruerus Holanus." Arguably, then, reading the works of "Jordanus Brunus Nolanus" would have been no crime in Iberia, but it is likely that no one dared to find out for certain. Furthermore, the elaborate mechanism by which books were first censored and then approved for publication (with the Latin term *Imprimatur*—"let it be printed") made the printing of books in Catholic countries a cumbersome and risky procedure. The Protestant presses of Northern Europe took no time at all to exploit their advantage, so that even Catholic scholars in Italy, like Kircher himself, eventually entrusted much of their work to Dutch and German printers.

Rules for censors, from the Lisbon *Index* (1624):

> Whosoever takes up the business of correcting and expurgating ought to observe everything and take careful note not only of the matters that manifestly offer themselves in the course of the work, but also whatever lies hidden, as if in ambush, in the notes, the summaries, the marginalia, and indices of books, in prefaces, and letters of dedication. (P. B2 verso)

Athanasius Kircher did not escape the attention of the Inquisition. A report on the censorship of his *Ecstatic Heavenly Journey* (*Iter exstaticum*), filed with his Jesuit superiors in Rome, explicitly addresses the conflict between his writings and Church doctrine:

> Kircher says beyond a doubt that the fixed stars are not located on the same surface of the firmament, or at the same remove from the earth, but that they are contained in intimate, immense recesses of that same firmament, removed at an inexplicable distance, and indeed that they are of a numberless multitude, so that the number of stars exceeds the number of the Elect. (Censorship of the *Iter exstaticum*, Biblioteca Nazionale Centrale, Rome, Fondo Gesuitico 1331, fasc. 15, p. 207 verso)

> To be sure, Kircher on occasion reproves the condemned opinion of Copernicus about the motion of the earth, lest (he says on p. 28) he be seen to assert anything contrary to the decrees and institutions of the Holy Roman Church: nonetheless, throughout his entire book he carefully constructs all the evidence that Copernicus first brought in to establish and defend the motion of the Earth, and he weakens all the arguments by which that error is usually refuted under a great weight of reasoning. From whom, if not from Copernicus and his followers, did Kircher accept that immensity of the firmament that he inculcates ad nauseam, and that error about the distance of the fixed stars from the earth? (Ibid., p. 209 recto)

> Finally, he may reject the motion of the earth on pages 73 and 320, and impugn it, but he does it so poorly, as we shall show . . . he is evidently prevaricating on the matter, and he is not doing so from the heart, but in order not to say anything openly contrary to the decrees and institutions of the Holy Roman Church. But he would have been more obedient if he were to deny his opinions, which he does in passing and in word only, beyond every probable reason. . . . (Ibid., pp. 210 recto–verso)

> For if the primary qualities are not accidental, then the secondary qualities that arise from the primary will also not be accidental, but just as the primary bodies are a kind of effluvium, then the secondary are made by mixing such effluvia. Hence it is manifest that there will be no accidentals in nature. But this cannot stand according to the doctrine of transubstantiation in the Eucharistic sacrament, as is evident enough without further explanation. Therefore Kircher's teaching is plainly extremely dangerous to doctrine. (Ibid., pp. 222 verso–223 recto)

107. *Index librorum prohibitorum.* Rome: Ex typographia Rev. Cam. Apost., 1667. Bound with *Index librorum prohibitorum.* Madrid: Ex typographæo Didaci Diaz, 1667.
Berlin Collection

This edition of the *Index of Prohibited Books* was issued in 1667 under the auspices of the former inquisitor Pope Alexander VII. The list of forbidden books for 1633, shown here, includes the name of Melchior Inchofer, whose book on a purported letter of the Virgin Mary to the citizens of Messina in Sicily was deemed "in need of correction." At the same time, Inchofer was working for the Inquisition as an examiner of Galileo's *Dialogue on the Two Chief World Systems*, which is forbidden in the papal bull of 1634, printed on the facing page.

108. *Index auctorum damnatae memoriae.* Lisbon: Ex officina Petri Craesbeeck, 1624.
John Crerar Collection of Rare Books in the History of Science and Medicine

Even Dante's *Divine Comedy* ran afoul of the Spanish Inquisition. When this copy of the *Index* was issued in Lisbon in 1624, Portugal had been annexed by Spain, and would not win independence until 1640; therefore the Spanish Inquisition dictated its contents.

109. *Indicis librorum expurgandorum.* Bergamo: Typis Comini Venturæ, 1608.
Berlin Collection

When this copy of the Roman *Index* was printed in Bergamo in 1608, Giordano Bruno's death at the stake in the Eternal City was still a vivid memory, and the bull prohibiting his writings still fresh. The Inquisition had never been quite certain exactly what Bruno wrote, and hence the bull makes a generic prohibition of "all written work."

110. Konrad Gesner (1516–1565). *Historiæ animalium.* Vol. 3. Frankfurt: Ex officina typographica Ioannis Wecheli, impensis Roberti Cambieri, 1585.
John Crerar Collection of Rare Books in the History of Science and Medicine

The Library's set of the multivolume *History of Animals* (*Historiae animalium*) by the immensely prolific Swiss naturalist Konrad Gesner is a mixed one, with this 1603 edition of volume 1; a 1587 edition of volume 5, also shown here; and other volumes, ranging from 1585 to 1604. All, however, were censored at the same time and then made suitable for reading, or "permitted with expurgation" (*cum expurgatione permissus*): all complimentary references to Protestant or heretical scholars—mostly adjectives like "learned"—have been deleted. According to proper inquisitorial practice, the censor, a Spanish Jesuit named Juan de Pineda (1558–1637), has signed his name on each title page of the work's several volumes. As the author of a treatise *On Christian Agriculture*, Pineda was presumably especially well-qualified to assess the works of his famous Swiss counterpart.

111. Konrad Gesner (1516–1565). *Historiæ animalium.* Vol. 5. Zurich: In officina Proschouiana, 1587.
John Crerar Collection of Rare Books in the History of Science and Medicine

In this expurgated page of Gesner's *History of Animals*, the text that has been removed by Pineda describes a Protestant scholar as "learned."

CAT. 111

112. Robert Cardinal Bellarmine, S.J. (1542–1621). *Riposta del Card. Bellarmino a due libretti.* Rome: Appresso Guglielmo Facciotto, 1606.

The Italian Jesuit theologian Robert Bellarmine (Roberto Bellarmino) was one of the most authoritative voices for the doctrinaire Church. A brilliant thinker, he is best known for his *Controversies*, a series of doctrinal arguments made with precision and impeccable logic. As cardinal, he acted as one of Giordano Bruno's examiners and was the person who brought Bruno's eight-year trial to its tragic conclusion. Apparently chastened by the experience, he was more lenient with Galileo in 1616. Originally he had been interested in science himself, but ultimately bent his own will to what he perceived as the will of the Church. He was canonized in 1935; however, his native city of Montepulciano boasts a "Via Giordano Bruno" among its streets and gives Bellarmine nary a mention. This book contains two small polemical tracts, against the theologian Giovanni Marsili and the historian Paolo Sarpi. The page shown reveals Bellarmine's anti-Semitism; it also features a name whose connection with the Spanish Inquisition is legendary: Torquemada, although Bellarmine referred to the Dominican theologian Juan de Torquemada rather than to his cousin Tomás, who founded the Inquisition in Spain and rendered it an instrument of terror.

> Saint Paul said, "I am at Caesar's tribunal; I appeal to Caesar." Cardinal Torquemada responds well to this argument, which was proposed by certain heretics long ago: that Saint Paul was forced to appeal to Caesar and recognize him as his de facto judge, not de jure, because at that time the power of Saint Peter was neither recognized nor credited, and therefore, if Saint Paul had said that he recognized no other judge than the Vicar of Christ, he would have made everyone laugh, both the Jews by whom he was accused and the Gentiles by whom he was judged. (p. 74)

# Panspermia, *the Power of the Universal Seed*

FOR KIRCHER, NATURE maintained its permanence through constant change, a cycle of growth and decay.

> In His divine and ineffable wisdom the All-Wise Creator took care that generation would follow corruption, and new corruption would succeed generation, according to the nature granted each individual creature by the Maker of that same universe, and thus the universe abides in its perfection through the wondrous vicissitude of things following upon one another in turn. This was nothing other than a seminal or spermatic power, by whose qualities and efficacy things appear by natural propagation from what has perished. (Athanasius Kircher, *Mundus subterraneus* [1678], 2.347)

He believed, following Aristotle, that the world was pervaded by this universal principle of fertility, which he often called *panspermia*, Greek for "universal seed." Aristotle recognized both male and female types of seed, although he declared that the male was by far the stronger. Kircher's seeds are also male and female; what connects them is magnetism, which he regarded as a physical expression of God's love:

> And just as in rational nature no affection of the passions is stronger than *philomania*, or the perturbation of Love, thus among the stones none has nearer and closer effects than this one. For who has not seen how intensely it is captured by love's firestorm in the presence of iron, how knowingly it pursues its beloved with motions, up and down, until its desired embrace, to obtain which it is carried headlong toward its natural center, nor will it desist from its lust for the iron in its presence until, querulous and agitated by affection until it has united with it, it finally finds its rest at its natural pole, and just as the human heart is visibly captured by the urgings of an inundation of love without visible chains, so too this stone, captured by its love for its desired iron is penetrated by the ties of an invisible embrace, so that they cannot be separated from one another except by force. (Filippo Buonanni, *Musaeum Kircherianum* [1709], p. 18)

> The very plantlets that lie buried in the womb of their seeds, under the gaze of the Sun, spring forth dilated with joy and soon bud off into leaves, flowers, fruits. All the animals, incited by the joy of the heavens, that is, the fertile radiation of light, as if by laughter, are stimulated to pleasure by fertilizing movement. The very rocks, remote as they may seem from every contact with light, attracted by some hidden power of radiation swell and in their tumescence rush to embrace each other, all joining the dance of the heavenly spheres. (Athanasius Kircher, *Ars magna lucis et umbrae* [1646], p. A verso)

CAT. 113

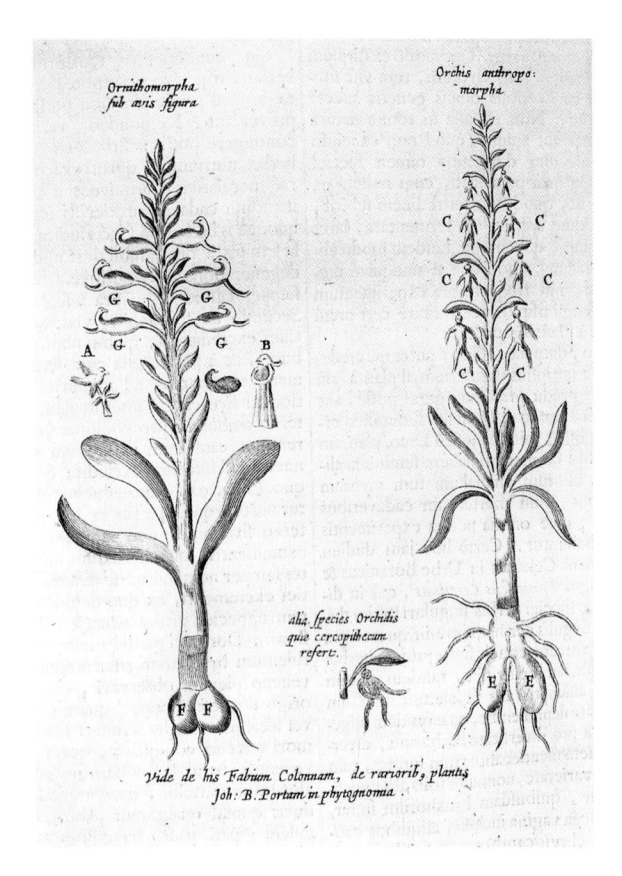

113. Photograph reproduced from Athanasius Kircher (1602–1680). *Mundi subterranei tomus II*. Vol. 2 of *Mundus subterraneus, in XII libros digestus.*
Amsterdam: Typis Joannis Jansonij à Waesberge et Elizæi Weyerstraet, 1665.
John Crerar Collection of Rare Books in the History of Science and Medicine

The ancient Greeks gave orchids their name, using a word (*orchides*) that means "testicles" because of the plant's suggestively shaped bulbs.

> Many Botanists, when they contemplate the wonderful varieties of Orchids (Orchids are those plants that are commonly called Satyria) cannot stop marveling at Nature's power in producing them. For if you consider their roots, you will find that they are in the form of testicles, from which they get their name. They are a rare and elegant species of plant, among which there are several that are called, not inappropriately, Anthropomorphic—Nature has played a game, in which there is hardly a part of the body that she has not tried to imitate to the extent that she can. (PP. 348–49)

114. Athanasius Kircher (1602–1680). *Magnes sive de arte magnetica*. Rome: Typis Vitalis Mascardi, 1654.
From the Library of Joseph Halle Schaffner

The ancient Greeks believed that the "testicular" shape of orchid bulbs conferred powers of fertility, an idea to which Kircher still subscribed in the seventeenth century; he believed that the extravagant flowers of the orchid resembled animals because the plants themselves and the soil around them had been fertilized by animals. No plant, however, could create a complete animal, but only a hybrid such as an animal-shaped flower.

> Just as every part of an animal is present in its seed by its very nature, as we said earlier, and they evolve bit by bit into the perfect form of an animal in the heat of the womb, with formative power acting to shape the individual parts, is it any wonder, I say, if the seed of men, or animals, when received into the harmonious, proportioned matrix of moist earth should end up forming, together with seed's own property to stimulate eros, if not a perfect animal, which would be beyond its powers, at least something analogous, whether to an animal or a man, whose parts are not greatly unlike the original? (Athanasius Kircher, *Mundus subterraneus* [1678], 2.369)

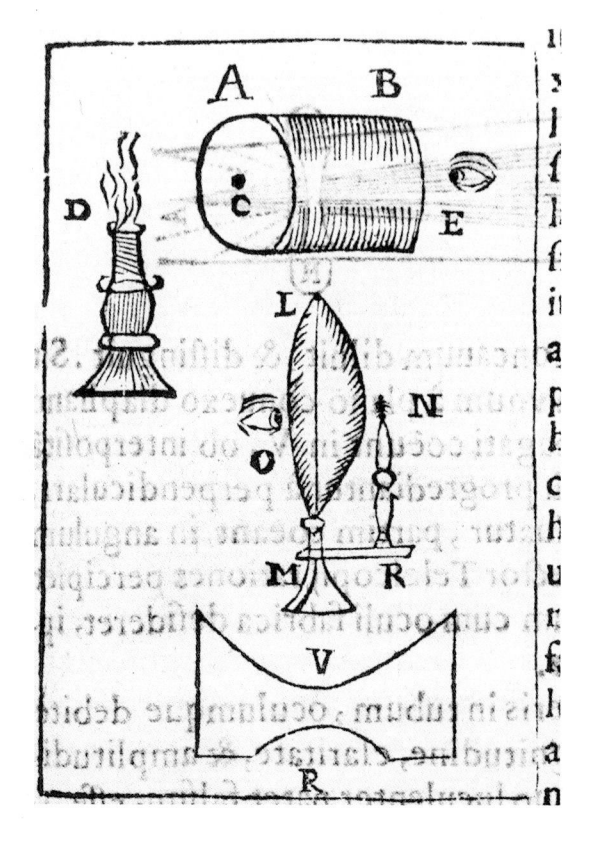

CAT. 115

104

115. Photograph reproduced from Athanasius Kircher (1602–1680). *Ars magna lucis et vmbrae*. Rome: Sumptibus Hermanni Scheus; Ex typographia Ludouici Grignani, 1646.

This woodcut of Kircher's *Smicroscopium* shows a rudimentary instrument with a single lens. Soon, however, like many of his fellow natural philosophers, Kircher began to make and use compound microscopes, with multiple lenses of far greater magnifying power. Like the telescope, the invention opened up a whole new world:

> Who could have believed that vinegar and milk teem with a numberless multitude of worms, unless the Smicroscopic Art had taught us so in these recent times, to the great wonderment of all? Who could ever have been made to think that the green color in citron leaves was composed of every kind of color, unless the Smicroscopic Art had revealed it? All these things experiment (literally, "experience"), the indomitable mistress of things, teaches us. (P. 834)

116. Athansius Kircher (1602–1680). *Scrutinium physico-medicum*. Rome: Typis Mascardi, 1658.

Written in the aftermath of the plague of 1655, this *Physico-medical Examination of the Contagious Pestilence Called the Plague* (*Scrutinium physico-medicum*), dedicated to Pope Alexander VII, ascribed the disease to a microscopic agent; it was one of the very first books to do so. Kircher called all microbes "worms" because they wiggled under the *Smicroscopium*:

> These worms that are the propagators of plague are so tiny, so slender and subtle, that they elude the senses' every power of comprehension: unless they were visible under the most finely tuned *Smicroscopium* you would call them atoms . . . they move about like atoms when sunlight is projected into a dark corner. (P. 141)

Eventually Kircher's habit of putting everything he could under the lens of his *Smicroscopium* caused a change in his eating habits:

> About fruits, it has to be said that they are more dangerous the more they are susceptible to decomposition. Hence the immoderate consumption of fruits that propagate worms should be avoided at top speed, as well as fruits with pernicious juice, like peaches, melons, watermelons; their like is to be avoided. Now the consumption of mushrooms, which are living pouches of poison, ought to be abolished altogether. (P. 28; cf. P. 122 on pernicious fungi)

If you examine the powder of rotten wood under the *Smicroscopium*, an immense pullulation of little worms will be found, of which some are outfitted with little horns, some have wings of a sort, others are not unlike centipedes, and you will see eyes like little black dots along with noses; thus God, Greatest and Best, shows Himself as marvelous not only in the vast bodies of the world, but also in the tiniest little animals, imperceptible to every eye, when he gave them their individual parts, without which they could not move nor perform any of their other vital acts. Because they themselves have been placed in the world with bodies so tiny that they are beyond the reach of the senses, how tiny can their little hearts be? How tiny must their little livers be, or their little stomachs, their cartilage and little nerves, their means of locomotion? (P. 45)

117. Photograph reproduced from Athanasius Kircher (1602–1680). *Mundi subterranei tomus II*. Vol. 2 of *Mundus subterraneus, in XII libros digestus*. Amsterdam: Typis Joannis Jansonij à Waesberge et Elizæi Weyerstraet, 1665.
John Crerar Collection of Rare Books in the History of Science and Medicine

The many-breasted deity whose statue appears in the near background of the frontispiece for the second volume of the *Subterranean World* (*Mundus subterraneus*) is the famous statue of Artemis, or Diana, whose huge Ionic temple at Ephesus figured as one of the Seven Wonders of the ancient world. Kircher and many of his contemporaries identified her as both an image of Isis, and, as here, of Mother Nature. They would probably be deeply disappointed to learn that recent excavations of the site at Ephesus have revealed that what looked like multiple breasts on ancient statues of the goddess were really huge amber beads shaped like teardrops, hung on the statue as votive offerings. The bustling port city was one of the largest in the ancient Mediterranean, and the temple one of the great tourist attractions of antiquity: Saint Paul narrowly escaped a riot in the theater when he came to preach, for local souvenir-makers panicked at the thought of a new religion carrying away their trade. Only a few centuries after Roman emperors had created a beautiful layout of colonnaded streets, Ephesus was abandoned; silt had clogged its once-magnificent harbor.

> *And when the townclerk had appeased the people, he said, Ye men of Ephesus, what man is there that knoweth not how that the city of the Ephesians is a worshipper of the great goddess Diana, and of the image which fell down from Jupiter?*
>
> (Acts 19:35)

ATHANASII KIRCHERI
E Soc. IESU
MUNDI SUBTERRANEI
TOMUS II.us
IN V. LIBROS DIGESTUS Quibus
Mundi Subterranei fructus exponuntur, et
quidquid tandem rarum, insolitum, et portentosum
in foecundo Naturæ utero continetur, ante oculos
ponitur curiosi Lectoris.

**Orpheus**
Ὅς ναίεις κατὰ πάντα μέρη κόσμοιο γενάρχα
Ὅς δαπανᾶς μὲν ἅπαντα, καὶ αὔξεις ἔμπαλιν αὖτες
Omnes qui partes habitas, mundique Genarcha
Absumis que cuncta eadem, quæ rursus adauges.

C. vande Pas delineavit.     F. Sioutsma Sculpsit.

AMSTELODAMI, Typis Joannis Janssonij à Waesberge et Elizæi Weyerstraet 1665.

CAT. 115

# *The Debate on Spontaneous Generation*

FATHER KIRCHER'S DOCTRINE of *panspermia*, the power of universal fertility, had its roots in Aristotle, including the theory that this universal spermatic power could bring about spontaneous generation of plants and animals. Animal seed deposited in an inadequate womb could not reproduce a real animal, but neither could its power of fertility go unexpressed; the result would be an imperfect animal like an insect, a mollusk, a worm, or perhaps a plant like an orchid. As Kircher stated in the *Subterranean World*:

> We have seen that there is no animal that does not of itself generate some other imperfect animal of a different species (just as we also said about plants). (2.372)

From 1668 onward, however, this theory of spontaneous generation came under increasing attack, especially by a group of Italian scholars who worked under the influence of Francesco Redi (1626–1698), the influential court physician to the grand duke of Tuscany. Redi countered the essentially masculine principle of Kircher's spermatic power with a theory that "lower" animals like insects came from eggs, bolstering his arguments with experimental evidence. As scientific instrumentation continued to improve, the debate hinged more and more on the design and interpretation of experiments rather than on principles of natural philosophy. Kircher's side of the argument was taken up after his death by his associate at the Collegio Romano, Father Filippo Buonanni (1638–1725), who applied rigorous logic and outstanding laboratory technique to the defense of Aristotle, and who was a sufficiently objective scientist to accept the experimental results that finally damaged the case for spontaneous generation beyond repair.

118. Francesco Redi (1626–1698). *Esperienze intorno alla generazione degli insetti*. Florence: All'insegna della stella, 1668.

The most vocal opposition to the theory of spontaneous generation came from the brilliant, hot-tempered, insufferably vain court physician to the grand duke of Tuscany, Francesco Redi of Arezzo, who made his first assault in this little book, *Experiments on the Generation of Insects* (*Esperienze intorno alla generazione degli insetti*). Here, Redi argued that insects came from larvae—"worms"—which had hatched from eggs deposited in plant galls, dung, and putrefying matter by adult insects. With lasting effect on modern scientific method, Redi also insisted strenuously that experiments could claim no validity unless their results could be reproduced. An admirer of Galileo and a leader of the short-lived but illustrious Tuscan scientific society, the Accademia del Cimento, Redi wrote, like his idol, in Tuscan vernacular, the diplomatic language of the Grand Duchy.

119. Francesco Redi (1626–1698). *Esperienze intorno a diverse cose naturali*. Florence: All'insegna della nave, 1671.
John Crerar Collection of Rare Books in the History of Science and Medicine

Francesco Redi addressed a second volume rejecting spontaneous generation directly to Kircher, as its title shows: *Experiments About Various Natural Phenomena . . . Written as a Letter to the Most Reverend Father Athanasius Kircher of the Society of Jesus* (*Esperienze intorno a diverse cose naturali. . . .*) Again he wrote in Tuscan vernacular rather than Latin, in part because Kircher himself corresponded as often, and as fluently, in Italian vernacular as he did in Latin, but largely out of Tuscan patriotism; in a sense, vernacular had come to symbolize scientific progress—at least in the minds of Galileo's fellow citizens.

120. Photograph reproduced from Francesco Redi (1626–1698). *Esperienze intorno alla generazione degli insetti*. Florence: All'insegna della stella, 1668.

Every user of the microscope was shocked to see tiny insects like fleas and lice enlarged to monstrous size, and artists like Redi's talented illustrator Filizio Pizzichi rose to the challenge of depicting them; the delicately painted, colored originals on which Pizzichi based his engravings are still preserved in the National Library in Florence.

Redi and his partisans directed diligent attention to the search for insect eggs as part of their continuing attack on the theory of spontaneous generation. They also examined fungi and mollusks in hopes that they would discover the eggs on which they based their own theories of generation.

121. Filippo Buonanni (1638–1725). *Ricreatione dell'occhio e della mente nell'osservation' delle chiocciole*. Vol. 2. Rome: per il Varese, 1681.
John Crerar Collection of Rare Books in the History of Science and Medicine

A close associate of Kircher's at the Collegio Romano, Father Filippo Buonanni took over the *Musaeum* after his mentor's death in 1680. He also took up the defense of the theory of spontaneous generation, beginning with this *Recreation of the Eye and Mind through Observation of Snails* (*Ricreatione dell'occhio e della mente nell'osservation' delle chiocciole*), whose life cycle he examined under the microscope. Writing in Tuscan, Buonanni clearly intended to address his remarks chiefly to Francesco Redi and his adherents. The fanciful assemblage of seashells shown here, reminiscent of seventeenth-century decorative pieces and modern souvenirs, has been arranged to recreate a classical battle trophy (compare the Trophies of Marius, cat. no. 6), the symbolic token of Buonanni's victory over Redi in the latest engagement of the great war over spontaneous generation. Buonanni's victory, alas, was a Pyrrhic one; Marcello Malpighi's observation of insect eggs would put a swift end to the debate. The caption says, "The Mollusk shells described in Part Two are portrayed here."

122. Filippo Buonanni (1638–1725). *Ricreatione dell'occhio e della mente nell'osservation' delle chiocciole.*
Vol. 1. Rome: per il Varese, 1681.
John Crerar Collection of Rare Books in the History of Science and Medicine

Filippo Buonanni based his arguments for spontaneous generation on experiments with snails, but his curiosity, nearly as broad as his teacher's, extended to all the mollusks. This *Recreation of the Eye and the Mind through Observation of Snails* is important for its attempt to classify the mollusks systematically, as well as for its close observations. Here Buonanni explained how snails move:

> Because shellfish have neither blood nor heart, to which their limbs are proportioned, they must contain the principle of motion in every part of their bodies, and therefore have many organic parts adapted to that purpose. . . . Most move by waves, as this is the most effective motion by which to move forward. (PP. 316–17)

123. Filippo Buonanni (1638–1725). *Observationes circa viventia, quae in rebus non viventibus reperiuntur.*
Rome: Typis Dominici Antonij Herculis, 1691 [i.e. 1692].
From the Library of Joseph Halle Schaffner

Father Buonanni's *Observations About Living Things That are Found in Non-Living Things* (*Observationes circa viventia, quae in rebus non viventibus reperiuntur*) is written in Latin, and was therefore clearly intended to address an international audience that reached as far as London's Royal Society.

124. John Carrera (b. 1969) and Sam Walker (1950–1999). *Putrefatti.* Boston: Quercus Press, 1995.
Number 17 of 25 copies.
Purchased on the R. R. Donnelley & Sons Company Fund

This artists' book takes its inspiration from Francesco Redi's pioneering *Experiments on the Generation of Insects* and from Filizio Pizzichi's stunning illustrations of insects observed under the microscope. Walker and Carrera make their own book disintegrate like one of Redi's experiments on putrefaction, as the type crawls away like larvae and the pages give birth to an enormous insect.

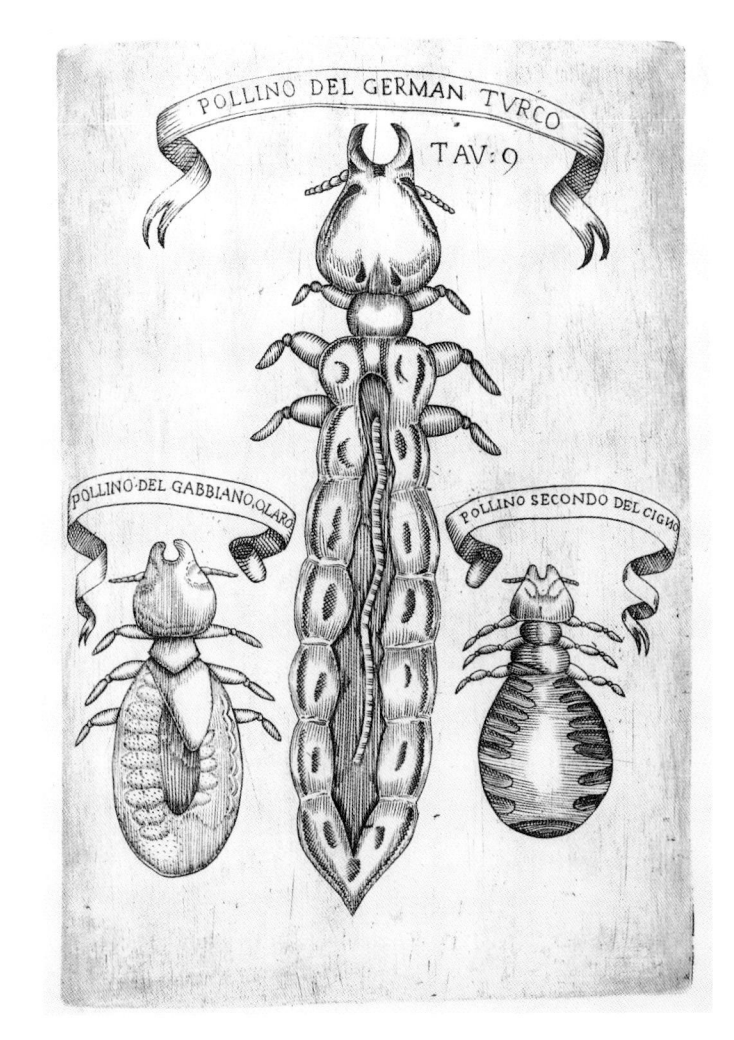

CAT. 120

3

4